Infectious Disease

CLINICAL CASES UNCOVERED

Infectious Disease

CLINICAL CASES UNCOVERED

Hamish McKenzie
PhD, FRCPath, FRCP (Ed), FHEA
Professor of Medical Education and
Honorary Consultant Microbiologist
School of Medicine
University of Aberdeen
Aberdeen, UK

Robert Laing
MD, FRCP (Ed)
Consultant Physician in Infectious Diseases and
Honorary Clinical Senior Lecturer
Aberdeen Royal Infirmary
Aberdeen, UK

Alexander Mackenzie
DTM&H, FRCP (Ed)
Consultant Physician in Infectious Diseases and
Honorary Clinical Senior Lecturer
Aberdeen Royal Infirmary
Aberdeen, UK

Pamela Molyneaux
FRCPath, FRCP (Ed)
Consultant Virologist and Honorary Clinical
Senior Lecturer
Aberdeen Royal Infirmary
Aberdeen, UK

Abhijit Bal
FRCPath
Consultant Microbiologist
Crosshouse Hospital
Kilmarnock, UK

WILEY-BLACKWELL
A John Wiley & Sons, Inc., Publication

Blackwell Publishing was acquired by John Wiley & Sons in February 2007. Blackwell's publishing program has been merged with Wiley's global Scientific, Technical and Medical business to form Wiley-Blackwell.

Registered office: John Wiley & Sons Ltd, The Atrium, Southern Gate, Chichester, West Sussex, PO19 8SQ, UK

Editorial offices: 9600 Garsington Road, Oxford, OX4 2DQ, UK
The Atrium, Southern Gate, Chichester, West Sussex, PO19 8SQ, UK
111 River Street, Hoboken, NJ 07030-5774, USA

For details of our global editorial offices, for customer services and for information about how to apply for permission to reuse the copyright material in this book please see our website at www.wiley.com/wiley-blackwell

Library of Congress Cataloging-in-Publication Data
Infectious disease / Hamish McKenzie . . . [et al.].
p. ; cm. – (Clinical cases uncovered)
Includes bibliographical references and index.
ISBN 978-1-4051-6891-5 (alk. paper)
1. Communicable diseases–Case studies. I. McKenzie, Hamish. II. Series: Clinical cases uncovered.
[DNLM: 1. Bacterial Infections–diagnosis–Case Reports. 2. Diagnosis, Differential–Case Reports. 3. Virus Diseases–diagnosis–Case Reports. WC 200 I43 2009]
RC112.I4575 2009
616.9–dc22 2009004442
ISBN: 978-1-4051-6891-5

A catalogue record for this book is available from the British Library.

Set in 9 on 12 pt Minion by SNP Best-set Typesetter Ltd., Hong Kong
Printed and bound in Singapore by Markono Print Media Pte Ltd

1 2009

Contents

Preface, vii

Acknowledgements, viii

How to use this book, ix

List of abbreviations, x

Part 1 Basics, 1

Laboratory diagnosis of infection, 1

Antimicrobial chemotherapy, 13

Infection and immunity, 22

Approach to the patient, 28

Part 2 Cases, 34

Case 1 A 68-year-old woman with bloody diarrhoea, 34

Case 2 A 73-year-old man who has been feeling generally unwell for 2 weeks, 38

Case 3 A 36-year-old man with hospital-acquired pneumonia, 42

Case 4 A 26-year-old male with jaundice, 45

Case 5 A 28-year-old female student with severe headache and drowsiness, 49

Case 6 A 30-year-old female with a vesicular skin rash, 57

Case 7 A 37-year-old man with fever and pleuritic pain, 61

Case 8 A 14-year-old school girl with a rash, 66

Case 9 A 68-year-old woman with fever and muscle pains, 70

Case 10 A 35-year-old teacher with fever and chills on return from Malawi, 74

Case 11 A 32-year-old man with night sweats and fatigue, 78

Case 12 A 41-year-old male with a red, hot and swollen left lower leg, 82

Case 13 An 18-year-old female student with vaginal discharge, 86

Case 14 An outbreak of diarrhoea and vomiting on an orthopaedic ward, 89

Case 15 An antenatal visit, 92

Case 16 Headache in a 35-year-old South African woman, 96

Case 17 A 64-year-old man with fever and rigors, 101

Case 18 A 42-year-man with fever, cough and myalgia, 106

Case 19 A 75-year-old man with a sore hip, 109

Case 20 A 24-year-old man with acute myelogenous leukaemia who has developed a fever, 112

Case 21 A 53-year-old man with fever, severe back pain and abdominal pain, 115

Case 22 A 20-year-old female student with a rash, fever, myalgia and diarrhoea, 119

Case 23 A 54-year-old man with a cough and night sweats, 122

Case 24 A 15-year-old boy with fever and a sore throat, 125

Part 3 Self-assessment, 128

MCQs, 128

EMQs, 133

SAQs, 137

Answers, 140

Index of cases by diagnosis or organism, 145

Index, 147

Colour plate section can be found facing p. 20.

Preface

Infection is an exciting area of medicine for a whole variety of reasons. It combines the science of microbiology with the art of clinical practice, and it affects patients of all age groups in any or all of their bodily systems. The sense of detective work – each patient presenting a unique puzzle – is never far away and presents an intellectual challenge in addition to the clinical challenges of communicating with patients and dealing with ethical and highly personal issues. Healthcare-acquired infection has become an increasingly important problem and the combination of increasingly sophisticated healthcare with the natural evolution of new pathogens (or old ones with new virulence traits) means that there is a constant need for research and for doctors to keep up to date with current knowledge and practice. We hope that this book is a good starting point to stimulate your interest in infection. We have tried to make the sometimes complex scientific basis of infection relevant and understandable in the early chapters. The cases then provide examples of the range of clinical puzzles that infectious disease specialists and microbiologists try to solve on a daily basis. We hope that you will practise applying your knowledge as these cases unfold and that you will find this an enjoyable way of learning about infection. The self-assessment questions at the end will help you judge how much you have learned, but keep in mind that much of what you will need to know for the rest of your professional life has still to be discovered!

Hamish McKenzie
Robert Laing
Alexander Mackenzie
Pamela Molyneaux
Abhijit Bal

Acknowledgements

We are grateful to all our colleagues who helped with this textbook, particularly in providing X-rays and illustrations. Our particular thanks goes to staff in Medical Microbiology for their help with laboratory pictures and to our ever helpful Department of Medical Illustration in the College of Life Sciences and Medicine at Aberdeen University who provided excellent support in producing photographs and figures. Ms Jacqui Morrison provided excellent secretarial and administrative help. Lastly, we thank our spouses and families for their patience at times when they would rather have had us do things other than writing this book!

How to use this book

Clinical Cases Uncovered (CCU) books are carefully designed to help supplement your clinical experience and assist with refreshing your memory when revising. Each book is divided into three sections: Part 1, Basics; Part 2, Cases; and Part 3, Self-assessment.

Part 1 gives a quick reminder of the basic science, history and examination, and key diagnoses in the area. Part 2 contains many of the clinical presentations you would expect to see on the wards or crop up in exams, with questions and answers leading you through each case. New information, such as test results, is revealed as events unfold and each case concludes with a handy case summary explaining the key points. Part 3 allows you to test your learning with several question styles (MCQs, EMQs and SAQs), each with a strong clinical focus.

Whether reading individually or working as part of a group, we hope you will enjoy using your CCU book. If you have any recommendations on how we could improve the series, please do let us know by contacting us at: medstudentuk@oxon.blackwellpublishing.com.

Disclaimer

CCU patients are designed to reflect real life, with their own reports of symptoms and concerns. Please note that all names used are entirely fictitious and any similarity to patients, alive or dead, is coincidental.

List of abbreviations

AAFB acid and alcohol fast bacilli
A&E accident and emergency
AIDS acquired immune deficiency syndrome
AML acute myelogenous leukaemia
ARDS adult respiratory distress syndrome
ASO anti-streptolysin O
AZT zidovudine
BAL bronchoalveolar lavage
BCG Bacille Calmette-Guérin
bd twice a day
BIS British Infection Society
BP blood pressure
BV bacterial vaginosis
CDAD *Clostridium difficile*-associated diarrhoea
cDNA complementary DNA
CFT complement fixation test
CMV cytomegalovirus
CO_2 carbon dioxide
COPD chronic obstructive pulmonary disease
CPE cytopathic effect
CRP C-reactive protein
CSF cerebrospinal fluid
CT computerised tomography
CVP central venous pressure
CVS cardiovascular system
DIC disseminated intravascular coagulation
DNA deoxyribonucleic acid
DOTS directly observed treatment
EBV Epstein–Barr virus
ECG electrocardiogram
EHEC enterohaemorrhagic *Escherichia coli*
EIA enzyme immunoassay
EM electron microscopy
EPP exposure prone procedure
ERCP endoscopic retrograde cholangiopancreatography
ESBL extended spectrum β-lactamase
ESR erythrocyte sedimentation rate
FUO fever of unknown origin
FVU first void urine
GCS Glasgow Coma Scale
GI gastrointestinal
GP general practitioner
HBIG hepatitis B immunoglobulin
HCV hepatitis C virus
HDU high dependency unit
Hib *Haemophilus influenzae* type b
HIV human immunodeficiency virus
HNIG human normal immunoglobulin
HSV herpes simplex virus
HUS haemolytic uraemic syndrome
ICP intracranial pressure
ICU intensive care unit
IF immunofluorescence
IFN interferon
Ig immunoglobulin
IL interleukin
IV intravenous
LA latex agglutination
LDH lactate dehydrogenase
LP lumbar puncture
LPS lipopolysaccharide
MBC minimum bactericidal concentration
M,C&S microscopy, culture and antibiotic sensitivity testing
MDR multidrug resistant
MIC minimum inhibitory concentration
MLST multilocus sequence typing
MMR measles, mumps and rubella
MpV metapneumovirus
MRCP magnetic resonance cholangiopancreatography
MRI magnetic resonance imaging
MRSA methicillin-resistant *Staphylococcus aureus*
MSU mid-stream urine
NAAT nucleic acid amplification test

NICE National Institute for Health and Clinical Excellence
NO nitric oxide
OPAT out-patient antibiotic therapy
PBP penicillin-binding protein
P,C&O parasites, cysts and ova
PCP *Pneumocystis carinii* (now *jirovecii*) pneumonia
PCR polymerase chain reaction
PIV parainfluenza virus
PVL Panton–Valentine leukocidin
QD quinupristin-dalfopristin
RBC red blood cell
RNA ribonucleic acid
RSV respiratory syncytial virus
RT reverse transcriptase
SAH subarachnoid haemorrhage
SDA strand displacement amplification
SIADH syndrome of inappropriate antidiuretic hormone secretion
SIRS systemic inflammatory response syndrome
TAC transient aplastic crisis
TB tuberculosis
tds three times a day
TNF tumour necrosis factor
TOE transoesophageal echocardiogram
TSS toxic shock syndrome
TSST toxic shock syndrome toxin
TTE transthoracic echocardiogram
URTI upper respiratory tract infection
UTI urinary tract infection
VRE vancomycin-resistant enterococcus
VTEC verocytotoxic *Escherichia coli*
VZIG varicella zoster immunoglobulin
VZV varicella zoster virus
WBC white blood cell
WHO World Health Organisation
XDR extremely drug resistant
ZN Ziehl–Neelsen

Laboratory diagnosis of infection

There are three stages that might be considered in the process of arriving at a laboratory diagnosis of infection:

1 The selection of tests that are relevant given the clinical context.
2 The technical performance of these tests with appropriate quality controls.
3 Interpretation of the clinical significance of the results and decisions on appropriate management.

For most doctors, only outline knowledge of the second of these stages is needed, sufficient to enable them to deal with first and last effectively, at least for common infections. This chapter will attempt to cover all three stages and provide sufficient background for the reader to engage with the cases that follow later in the book. Approaches to the laboratory diagnosis of infection can be summarised as:

- Microscopic visualisation of the organism.
- Culture of the organism.
- Detection of organism-related antigen, toxin or nucleic acid.
- Measurement of the serological (i.e. antibody) response to an organism.

These will all be described in this section. There are two simple but important steps that the requesting clinician can take to make the process of laboratory diagnosis effective. The first is clear and accurate identification of the patient on the request form – most diagnostic microbiology laboratories process hundreds of thousands of specimens each year and proper identification is essential if an accurate report is to find its way back to the correct ward or practice. Secondly, the specimen type should be made clear and accompanied by a clinical history that will help laboratory staff to select appropriate tests. For a symptomatic patient, summarise the nature and duration of the illness and include the reason for taking the specimen. The reason for taking the specimen may not always be obvious from the main presenting complaint and one of the authors clearly remembers puzzling over the possible reasons for submission of an eye swab from a patient described as having "torsion of the testis". If the test is not for immediate diagnostic purposes (e.g. infection control screens, tests of immunisation response, etc.), make this clear on the request form.

Laboratory diagnosis requires a combination of scientific and technical expertise in processing clinical specimens, but there is also a major interpretive component in deciding on the most appropriate tests for each specimen and especially in weighing up the clinical significance of results. The interpretive component requires clinical judgement and is a major part of the role of medical microbiologists and virologists. Thus the diagnostic laboratory is more than a simple technical service and provides clinical advice on the diagnosis, management and prevention of infection.

The specimen journey: bacterial infection

The commonest sequence of events for a specimen collected for the diagnosis of bacterial infection is microscopy, culture and antibiotic sensitivity testing of any potentially significant organisms grown, often shortened to M,C&S on the request form. There is an important difference between specimens from a normally sterile site (e.g. blood, cerebrospinal fluid (CSF)) and those from sites with extensive normal flora (e.g. faeces, throat) when it comes to interpreting the results. Laboratories normally report only bacterial growth that is considered to be of clinical significance.

Infectious Disease: Clinical Cases Uncovered. By H. McKenzie, R. Laing, A. Mackenzie, P. Molyneaux and A. Bal. Published 2009 by Blackwell Publishing. ISBN 978-1-4051-6891-5.

Microscopy

Gram stain

The simplest and most rapid test for the presence of bacteria in a clinical specimen is microscopy. Most specimens are examined under the microscope after a **Gram stain**, which subdivides most bacteria on the basis of colour as Gram positive (purple) or Gram negative (red) and also on the basis of shape as bacilli (rod shaped) or cocci (spherical) (Plate 1). In order to visualise bacteria in a clinical specimen, they need to be present in very large numbers, so a negative result by microscopy does not rule out the presence of small numbers of bacteria in a sample. In addition, there are some bacteria that do not show up on a Gram stain, e.g. mycobacteria, *Legionella, Chlamydia*. It is not possible to identify a bacterial species fully on the basis of a Gram stain; for example, a Gram stain on a throat swab could not differentiate viridans streptococci from *Streptococcus pyogenes* and thus would not be of much clinical benefit. The Gram stain is most useful when it demonstrates the presence of an organism in a normally sterile site (e.g. CSF) but is less useful or more difficult to interpret if the specimen is from a site with a pre-existing normal flora (e.g. skin swab, faeces).

Microscopy of unstained specimens

This is most commonly used for urines and organisms can be readily seen (but not identified), as can pus cells (pyuria), red cells (haematuria) and casts. Increasingly laboratories are using automated technology (flow cytometry) to detect cells and organisms in urine.

Microscopy of unstained CSF is important in performing a cell count. Red cells and white cells are easily differentiated, and this is usually done in conjunction with a differential stain on a separate sample, which allows the proportion of neutrophils and lymphocytes to be estimated.

Auramine phenol and Ziehl–Neelsen stain

Mycobacteria are not visualised with the Gram stain and have traditionally been sought by microscopy using the Ziehl–Neelsen (ZN) stain (Plate 2). This stain involves a washing step with concentrated acid and concentrated alcohol, and thus positive organisms are described as acid and alcohol fast bacilli (AAFB). The auramine phenol stain is an alternative which is now more widely used for screening purposes as it is more sensitive. It must be viewed with ultraviolet illumination under the microscope and the ZN is generally used for confirmation of positives. As with Gram staining, large numbers of organisms must be present before microscopy is positive. It is not possible to identify the species of mycobacterium by microscopic examination, e.g. the laboratory cannot confirm that an organism is *Mycobacterium tuberculosis* and not an environmental mycobacterium on the basis of an auramine phenol or ZN stain.

Culture and sensitivity

After microscopy, clinical specimens are 'plated out' on agar plates (individual bacteria are distributed over the surface of the plate using a wire or disposable plastic loop) and then incubated. This pattern of plate inoculation and spreading (Plate 3) helps to separate complex mixtures of bacterial species in a clinical specimen. There is a wide choice of agar media on which specimens may be cultured and the selection is made on the basis of the organisms being sought. Thus a different range of plates would be used for a faeces specimen and a sputum specimen, as different pathogens are being sought. Incubation conditions can also vary depending on the organisms being sought. Culture plates from most clinical specimens will be incubated in air in the presence of 5% CO_2 overnight at 37°C. For most specimens, at least one plate will also be incubated anaerobically (in the complete absence of oxygen) to detect the presence of anaerobic bacteria. Although many organisms grow within 24–48 hours, a number require extended culture beyond 48 hours (frequently the case for anaerobes), while *Mycobacterium tuberculosis* can take several weeks to grow.

After culture, the next stage is the identification of any bacteria that have grown. Different species of bacteria produce colonies of widely differing morphology on different agar media, and the experienced microbiologist can make a provisional judgement on what organisms are present by examining the plates with the naked eye. Additional biochemical or other tests may be performed to confirm identity. The emphasis in the routine laboratory is to identify specimens with no obvious pathogens at an early stage and report that result with no further analysis. Thus a wound or throat swab that grew normal flora for that site would be discarded at this stage. By contrast, extensive work may be done on organisms that are potentially relevant pathogens given the specimen type and clinical background. These would be fully identified and antibiotic susceptibility tests performed.

A simplified classification of common pathogens is shown for Gram-positive organisms in Fig. A. Gram-positive cocci can be subdivided on the basis of two simple phenotypic properties – the coagulase test for

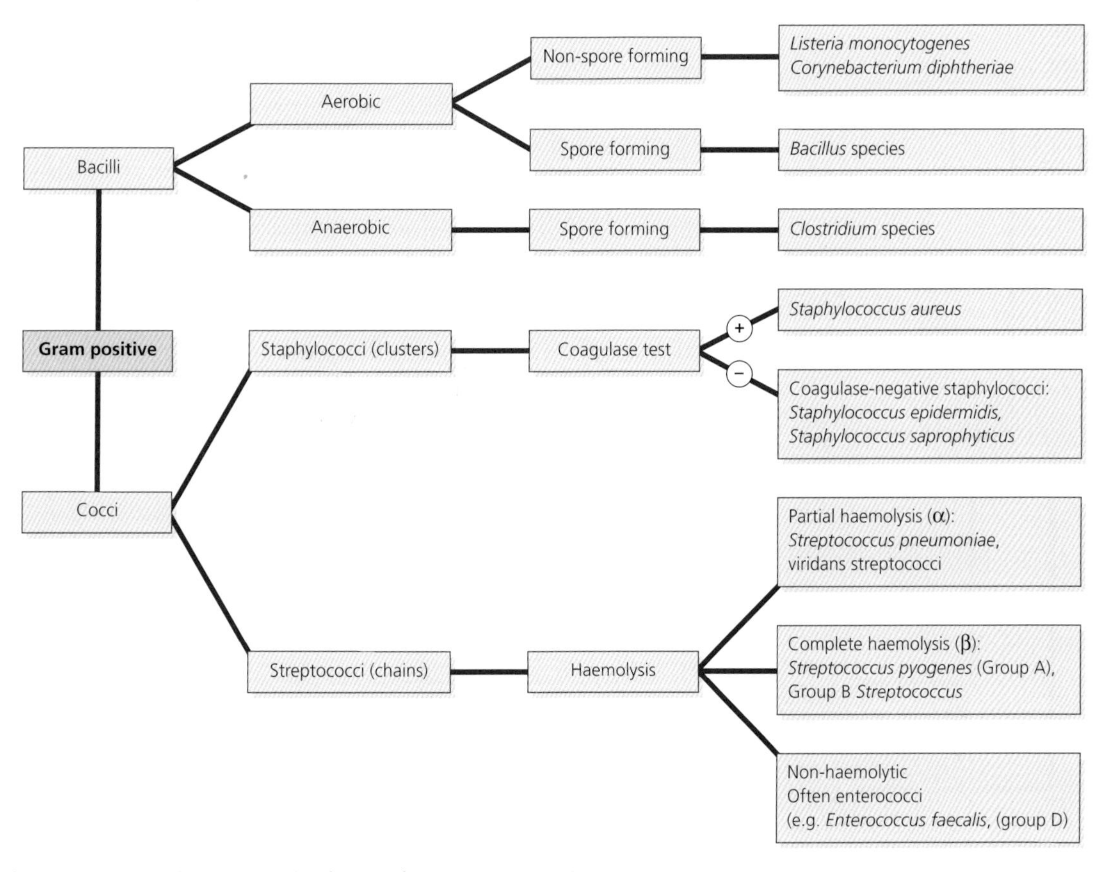

Figure A Flow chart showing basic classification of Gram-positive organisms.

staphylococci and the ability to haemolyse blood agar for streptococci. Examples of these tests are shown in Plates 4 and 5, respectively. *Staphylococcus aureus* is the only coagulase-positive staphylococcal species and is an important pathogen. There are many coagulase-negative species and these are usually of clinical importance only in infections of foreign bodies, e.g. prosthetic heart valves, prosthetic hip joints. Some streptococci (mostly β-haemolytic streptococci) possess surface antigens that allow further classification by **Lancefield grouping**. This is useful clinically and *Streptococcus pyogenes* (group A) is an important pathogen.

A simplified classification of common Gram-negative pathogens is shown in Fig. B. A wide range of Gram-negative bacilli, both aerobic and anaerobic, are found in the gastrointestinal tract. Members of the aerobic family Enterobacteriaceae are commonly called 'coliforms' as they are in the same family as *Escherichia coli*. Fermentation of lactose is a useful initial way of separating members of this group of organisms. The diarrhoeal pathogens *Salmonella* and *Shigella* are lactose non-fermenters, while *E. coli* and *Klebsiella* are lactose fermenters. *Pseudomonas* species are genetically different from colifoms and are notably more resistant to antibiotics.

Table A lists some common bacterial pathogens for which the Gram stain is not helpful as a means of detection or classification.

Antibiotic susceptibility tests

Organisms judged to be of clinical significance are tested for susceptibility against selected antibiotics, as discussed in Antimicrobial chemotherapy (Part 1). Limiting these tests to a small number of appropriate antibiotics is one of the ways in which antibiotic prescribing can be influenced and the development of resistance through inappropriate prescribing minimised. Note that *in vitro* tests can only give a guide as to whether antibiotic treatment will be effective in an individual patient with a particular infection. A whole variety of factors including dosage, route of administration and degree of penetration of the

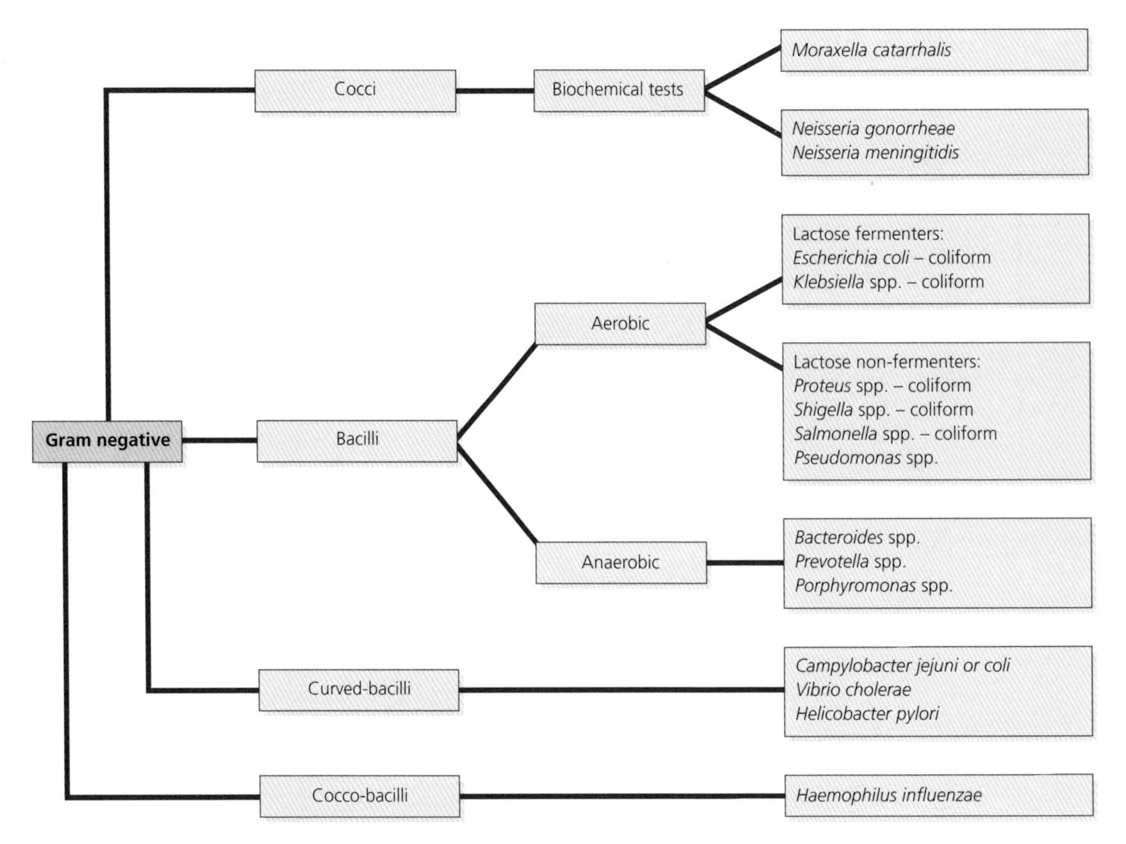

Figure B Flow chart showing basic classification of Gram-negative organisms.

Table A Common bacterial pathogens which are not routinely detected by Gram stain.

Mycobacteria	e.g. *Mycobacterium tuberculosis* (TB)
Spirochaetes	e.g. *Treponema pallidum* (syphilis), *Borrelia burgdorferi* (Lyme disease), Leptospira
Chlamydia	e.g. *Chlamydia trachomatis, C. psittaci*

antibiotic(s) to the site of infection will all influence outcome (see Antimicrobial chemotherapy, Part 1).

Antigen or toxin detection and bacterial infection

Bacterial antigens or toxins can be detected in various ways, but the commonest are enzyme immunoassay (EIA) and latex agglutination (LA). There is a more detailed technical explanation of how these tests work towards the end of this chapter. The commonest applications of these tests in the bacteriology laboratory are in the detection of *Clostridium difficile* toxin in stools (EIA), *Legionella* antigen in urine (EIA), pneumococcal, meningococcal and *Haemophilus influenzae* antigens in CSF (LA), and pneumococcal antigen in blood cultures or urine (LA).

Nucleic acid detection and bacteria

The polymerase chain reaction (PCR) revolutionised research in molecular biology by enabling scientists to amplify and detect known sequences of DNA. This approach is increasingly being applied to the diagnosis of infection. The detection of *Chlamydia trachomatis* DNA is the commonest use of this for bacterial diagnosis. Although *C. trachomatis* is a bacterium, testing has often been within the remit of the virology laboratory, since it cannot be cultured by conventional bacteriological techniques. Although commercial detection kits are available for other bacteria (e.g. for *Neisseria gonorrhoeae*), nucleic acid detection is not yet widely used in routine bacteriology laboratories because of costs and the availability of conventional culture. However, it is used in reference centres to characterise strains of clinically important bacteria and there is considerable interest in its possible use for the rapid detection of methicillin-resistant *Staphylococcus aureus* (MRSA) in clinical specimens. This would allow carriers to be identified on admission to hospital and then placed in isolation.

Serology and bacterial diagnosis

Serological testing allows a diagnosis to be made without culturing or detecting the organism itself. Instead, infection is indicated by detection of a specific antibody response to the pathogen concerned during the course of the illness. Thus this approach is usually employed when it is technically difficult to culture or detect the pathogen directly. Common bacterial infections for which serology is the major diagnostic approach include causative agents of the so-called 'atypical pneumonias' – *Chlamydia psittaci, Coxiella burnetii* and *Mycoplasma pneumoniae* – and spirochaete infections, e.g. syphilis (*Treponema pallidum*), Lyme disease (*Borrelia burgdorferi*) and leptospirosis. Interpretation of the results of serological testing requires an understanding of some basic immunological and methodological principles. These will be covered under 'Technical issues' later in this chapter.

The specimen journey: viral infection

Although there is some overlap, the mix of techniques commonly used in the diagnostic virology laboratory is different from the bacteriology laboratory and has changed considerably in recent years. Unlike bacteriology, microscopy and culture are of limited usefulness, and the emphasis is on antigen testing, nucleic acid detection and serology.

Microscopy and viral diagnosis

In the past electron microscopy (EM) was used widely in virus laboratories, but it has now largely been replaced (other than in a few reference centres) by other more sensitive test methods that are less demanding technically. EM requires a minimum of around 10^6 virus particles per millilitre of specimen suspension in order to have a reasonable chance of giving a positive result. The benefits of EM include the potential for a rapid diagnosis and the potential to see an unexpected organism.

Virus culture

As viruses are obligate intracellular parasites, they only grow within other living cells and therefore viral culture is more challenging technically than bacterial culture. Specimens for viral culture must therefore be cellular and be sent in viral transport medium, a preservative liquid that maintains viral viability and inhibits bacterial overgrowth. Living cells are inoculated with clinical specimens; any infecting virus present in the clinical specimen can infect the cells and then reproduce within them. After incubation, the infected cells may show changes, i.e. a cytopathic effect (CPE) that is visible on microscopy. Cell culture has a number of disadvantages: it is of low sensitivity, it requires viable virus in the sample, and it is technically demanding, slow (few viruses grow in 1–2 days and many take 2–3 weeks) and liable to contamination by bacteria, fungi and other viruses. Thus it is increasingly being replaced by nucleic acid detection.

Antigen detection and viral diagnosis

Antigen detection is a major component of the diagnostic virology laboratory's methodological approach. The commonest and most useful applications of antigen detection are detection of: (i) hepatitis B by EIA for surface antigen; (ii) a range of respiratory viruses, e.g. influenza A and B and respiratory syncytial virus (RSV) by immunofluorescence (IF); and (iii) detection of rotavirus in faeces by EIA or LA. For IF to be successful, cellular samples are essential as viral antigens are expressed on viral-infected cells (e.g. nasopharyngeal secretions are far preferable to a throat swab).

Nucleic acid detection and viral diagnosis

Nucleic acid detection is increasingly used for diagnosis of recent or current viral infections and because it is rapid, sensitive and does not require the presence of viable virus. It has largely replaced virus culture. It is possible to provide an estimate of the quantity of nucleic acid in a sample and thus to monitor viral load, e.g. of human immunodeficiency virus (HIV) during treatment, or of cytomegalovirus (CMV) or Epstein–Barr virus (EBV) in transplant recipients or other immunocompromised patients. This can be of clinical value in the diagnosis and treatment of such infections. For some infections, knowing the viral genotype (e.g. by nucleic acid sequencing) is needed for optimum patient management (e.g. in hepatitis C).

Serology and viral diagnosis

Serology is now used less frequently for the diagnosis of recent viral infection and more often for the detection of evidence of previous infection or successful immunisation. It remains important for the diagnosis of some acute viral infections, especially those for which there is an IgM assay available (see later in this chapter), e.g. hepatitis A, EBV, rubella and parvovirus (erythrovirus) B19.

The specimen journey: fungal infection

Microscopy and culture are the mainstays of diagnosis of fungal infection, with the morphological appearance of fungi under the microscope assuming much more importance in identification than it does in bacteriology. Fungi are found in two forms, as yeasts (unicellular) and moulds or filamentous fungi (multicellular strands that are called mycelia), with some species having both forms (dimorphic fungi) and others existing only as one or the other. Microscopy is an effective way of observing mycelia in clinical specimens – this would be commonly applied to skin scrapings (e.g. athlete's foot), nail clippings (e.g. onychomycosis) and sputum specimens (e.g. aspergillosis). The appearance of the mycelium, in terms of the branching patterns observed, provides an initial basis for identification of the species involved. Fungal culture requires specialised solid media and often takes several days. Identification of fungal growth on culture plates is done by macroscopic and microscopic appearance.

Yeasts are easily identified in Gram stains of clinical specimens and *Candida albicans* is the commonest species. Other species can cause disease and may have different antifungal susceptibilities, so the distinction can be clinically important. Detailed identification can be done by biochemical testing, but a simple laboratory test – the 'germ tube' – differentiates *C. albicans* from other species. As with bacteria, yeasts from sterile sites such as blood and CSF (e.g. candida meningitis in neonates or cryptococcal meningitis in immunosuppressed patients) are always significant, but yeasts found in non-sterile specimens are of uncertain significance at best.

Diagnosis by nucleic acid detection is not yet widely available for fungal infection other than for the respiratory pathogen *Pneumocystis jirovecii*, which causes infection in immunocompromised patients. *P. jirovecii* was previously classified as a protozoan and called *P. carinii*, but DNA studies have resulted in it being reclassified as a yeast. There have been many attempts to develop useful antigen detection tests for fungal infection (e.g. for invasive aspergillosis), but currently these are of limited value, the exception being the useful antigen test for confirmation of cryptococcal infection.

Antifungal susceptibility testing is not routinely available and presents many technical difficulties, although most laboratories do test *Candida* species for fluconazole susceptibility.

The specimen journey: parasitology

Parasites are largely detected by microscopy and their accurate identification requires skill and experience. Parasites are organisms that are dependent upon their host for survival and may be broadly classified into protozoa and helminths. They can infect any part of the human body but the commonest ones reside in the gut (e.g. tapeworms and pinworm). Others infect the brain (e.g. toxoplasma), heart (e.g. trypanosomes), liver (e.g. amoeba), muscles (e.g. trichinella), red blood cells (e.g. *Plasmodium* species, the malaria parasite) and macrophages (e.g. leishmania). The diagnostic approach to parasitic infections depends upon the type of parasite, but most commonly involves microscopy. Stool specimens are typically examined for 'parasites, cysts and ova' – often shortened on the request form to 'P,C&O please'. Many protozoa exist in an 'active' vegetative form and a 'resting' cyst form. Note that the vegetative form of *Entamoeba histolytica* can best be seen in a 'hot stool', freshly obtained and rushed to the laboratory. More commonly, amoebic cysts are seen in concentrated stools. Invasive amoebiasis is an important if uncommon (in the UK at least) cause of liver abscess and serological tests are useful for diagnosis. In the UK, *Giardia lamblia* and *Cryptosporidium parvum* are the two endemic protozoan infections most commonly diagnosed by detection of cysts in concentrated faeces. Malaria parasites are easily seen in blood films by experienced observers during an attack and malaria is an important diagnosis to exclude in the returning traveller. Serology is useful for certain parasitic infections, e.g. cerebral toxoplasmosis.

Parasitology is an important subspecialty within microbiology and its relative importance varies very much with geographical location and the patient population. In areas where parasitic infection is uncommon, expert advice (e.g. referral to a reference laboratory) is often required.

Technical issues

Much of the technical detail of laboratory tests is of importance only to those who work in the laboratory.

However, an understanding of the principles involved may help clinicians in the selection and interpretation of tests.

Antigen and toxin detection

The presence of an organism in a clinical specimen can be shown by the detection of specific antigen(s) or toxin(s). Three methods are commonly used for antigen detection: EIA, LA and IF. All of these methods use a monoclonal antibody for the desired antigen. Note that the normal human antibody response to a single organism involves the expansion of multiple clones of B cells into plasma cells to produce antibody of a wide range of specificities (i.e. polyclonal antibody). Monoclonal antibody is artificially produced from a single clone of plasma cells and therefore is a pure reagent with specificity for one single epitope on one single antigen (see Infection and immunity, Part 1 for details). Diagnostic kits to detect antigen or toxins are based on such reagents.

1 Latex agglutination. In LA, the monoclonal antibody is coated on latex particles and the cross-linking of antigen and antibody-coated particles produces a lattice of particles which is visible as clumping to the naked eye (Fig. C; Plate 6).

2 Immunofluorescence. In IF, the clinical specimen is smeared on a slide and fixed in position. The monoclonal antibody (which has a fluorescent dye bound to it) is applied to the slide and, after suitable incubation time to allow binding of the monoclonal to the antigen in question (if present), the excess fluid is washed off. IF staining can then be visualised under the microscope with ultraviolet light illumination (Fig. D; Plate 7).

3 Enzyme immunoassay. The exact design of EIAs can vary considerably. Many involve plastic strips of wells, each of which can hold a few hundred microlitres of liquid. Monoclonal antibody specific for the antigen is bound to the plastic surface (this is quite easy technically – the surface is sometimes referred to as the 'solid phase'). A clinical specimen in liquid form is added to the well and incubated. Any antigen present in the clinical specimen will be 'recognised' and bound by the monoclonal antibody during this incubation. Excess material is washed away from the solid phase and then a solution containing a second monoclonal, also specific for the antigen, is added. This second monoclonal has an enzyme bound to it and will become linked to the solid phase only if there is antigen present after the first incubation (Fig. E). After a further washing step, if there is enzyme-linked monoclonal bound to the solid phase, it is detected by adding a substrate for the enzyme. The substrate is selected to produce an easily measurable coloured product.

Nucleic acid detection

As the amount of nucleic acid present in a particular organism in a clinical specimen is too low to be detected directly, it needs to be 'multiplied' or amplified. Of the different nucleic acid amplification test methods available, the PCR is the most widely used. The more generic term nucleic acid amplification test (NAAT) is used to encompass a range of slightly different methodologies (see later in this chapter). NAAT requires knowledge of the nucleotide sequence of the target organism in question. The sequence chosen for amplification needs to be

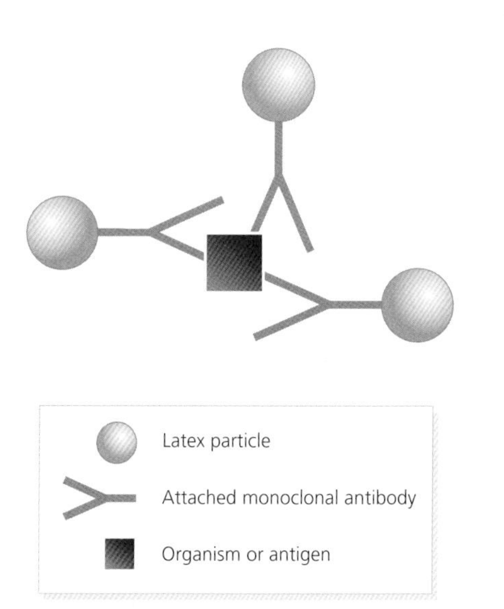

Figure C Diagrammatic representation of organism detection by cross linking of antibody-coated latex particles by antigen.

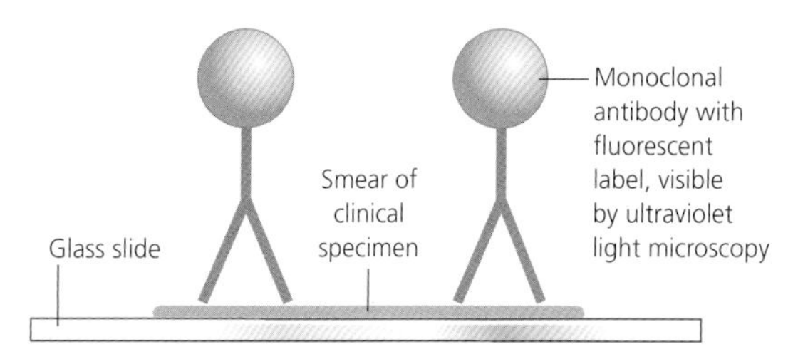

Figure D Diagrammatic representation of antigen detection by direct immunofluorescence.

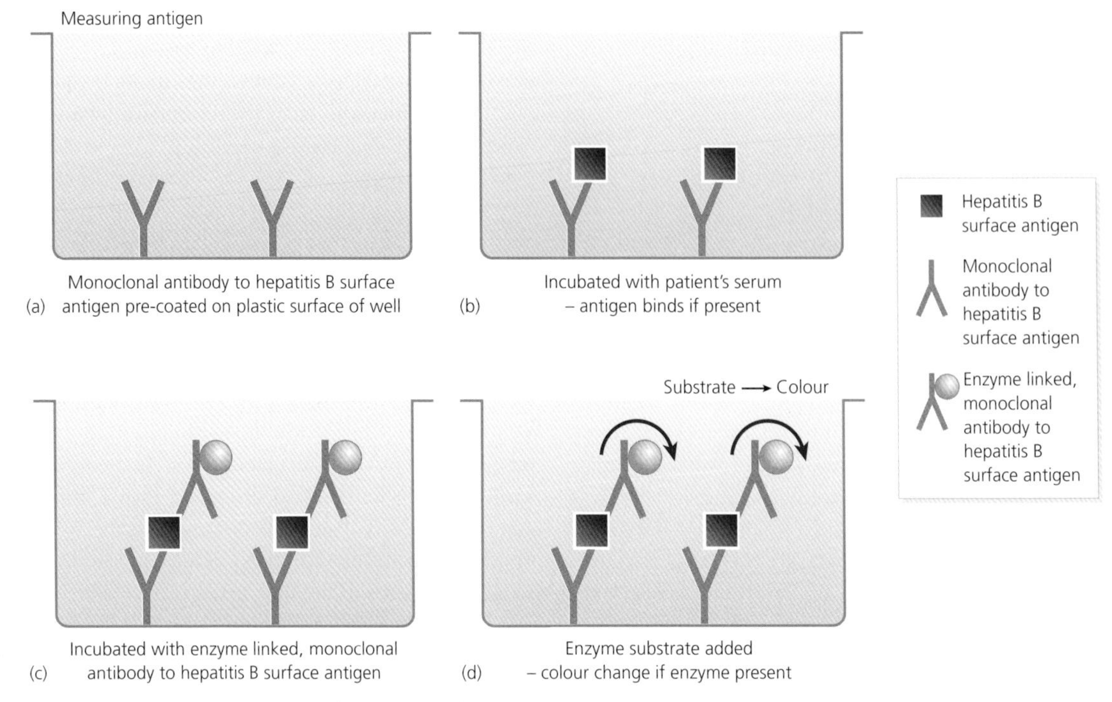

Figure E Diagrammatic representation of hepatitis B surface antigen detection by enzyme immunoassay.

unique for the target organism if specific organism detection is required, but in some cases NAAT may be used to detect the presence of a virulence gene of clinical relevance, e.g. a toxin gene. NAAT requires extraction of nucleic acid from the clinical sample followed by amplification and the detection of target nucleic acid sequence. Note that the sensitivity of PCR is such that very careful processing of samples is required at all stages (including sample collection!) to ensure that there is no cross contamination of nucleic acid, which could result in false positives. Real-time PCR (explained in more detail below) is a form of PCR that enables more rapid detection in a clinical sample.

Extraction

Firstly, nucleic acid is extracted from the specimen to be tested, with particular attention paid to the removal of potential inhibitors, e.g. enzymes such as nucleases that destroy DNA and RNA.

Amplification of specific sequence within the extracted nucleic acid by PCR

Amplification requires two different 'primers', short oligonucleotides usually 15–20 bases long, designed to bind to different strands of the target DNA at the 5′ ends of the target sequence. In PCR, double-stranded target DNA is first separated (denatured) by heating, and then the temperature reduced to allow binding (annealing) of the primers to their complementary sequence. An enzyme called Taq polymerase then allows nucleic acid synthesis to start from both primers (extension), incorporating nucleotides one by one to match the opposing strand (remember that adenine is incorporated opposite thymine (AT) and guanine opposite cytosine (GC)), thus producing two new complementary strands of DNA. The new strands of DNA are made in opposite directions, both from the 5′ end to the 3′ end (Fig. F). A series of heating and cooling 'cycles' (often about 40 cycles) are required to produce enough DNA for it to be detectable, with each cycle comprised of denaturing, annealing and extension. The test is carried out in blocks, which enables very rapid heating and cooling cycles such that 40 cycles can be completed in about 2 hours. If the target is RNA (e.g. to detect the presence of an RNA virus), a complementary DNA (cDNA) copy is made first, using the enzyme reverse transcriptase (RT), and this is then followed by the cycles of PCR, i.e. RT-PCR.

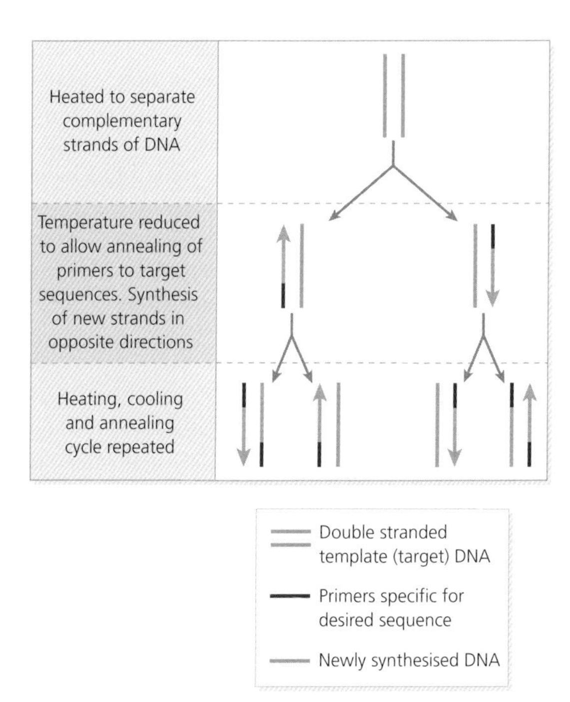

Figure F Diagrammatic representation of nucleic acid amplification by the polymerase chain reaction.

Detection of amplified nucleic acid

There are various ways in which the amplified DNA product can be detected at the end of the reaction, the exact details of which are not important here. The end point is a qualitative result (positive or negative) which should be validated by various positive and negative controls. Quantitation of the amount of target DNA present in the clinical sample is also possible and this is of obvious benefit in enabling 'viral load' to be determined in some infections (e.g. HIV). **Real-time PCR** uses a third oligonucleotide in the reaction mixture, a 'probe', which produces a fluorescent signal as the probe is increasingly incorporated in new DNA during synthesis. This fluorescence can be measured as the reaction proceeds, i.e. in 'real time' and allows rapid detection.

Other amplification systems

Since PCR is protected as an invention under patent, some manufacturers of diagnostic tests have developed other ways of amplifying target DNA molecules. Thus, for example, strand displacement amplification (SDA) uses an enzyme rather than a heating step to separate the two strands of DNA in each cycle.

Nucleic acid sequencing

Primers for NAAT are carefully selected for their specificity for the intended target, but also for areas of sequence that do not vary from strain to strain of the same species. However, sequence variation within a gene is sometimes useful as a means of differentiating between strains of a virus or of a bacterial species. Multilocus sequence typing (MLST) is an example of a commonly used bacterial typing method based on the sequences of (usually) seven 'housekeeping genes'. This enables strains within a species (e.g. a particular MRSA strain) to be identified and tracked, giving better epidemiological information on how the strain spreads and whether there are any particular patterns of infection associated with it. Sequencing techniques are the remit of specialist laboratories at present, but it is likely that they will become more widely accessible in the future and that sequence information on clinical isolates will become commonplace.

Serology

Serological tests are usually done on serum (separated from clotted blood) or plasma (separated from anticoagulated blood) and are designed to detect a specific antibody response to the pathogen concerned during the course of an infection. For serological diagnosis of a recent infection, the duration of the illness is important for the interpretation of results and should always be given on the accompanying request form. Measuring the antibody response instead of detecting the infecting organism is a logical, if indirect, approach to the diagnosis of infection, but there are disadvantages to this approach:

1 It can take some time for antibody to reach detectable levels and therefore serological diagnosis may not be possible at an early stage in an illness. It is important that a negative serology result is interpreted in the light of the duration of the illness, as a repeat sample may be indicated.
2 High levels of IgG antibody from previous infection or immunisation can remain in the circulation for years and it is sometimes difficult to distinguish 'old' antibody from the response to the current infection, e.g. in influenza.
3 Persistence or reappearance of IgM antibody, usually at a low level, may occur in some chronic infections, e.g. hepatitis B.

4 Serology is of limited use for diagnosing current or recent infection in immunocompromised patients due to their poor immune response (a false-negative result).
5 A non-specific reaction to a different, non-infecting organism can give a positive serological test result in the absence of that infection (a false-positive result).

Many serological tests are routine screening tests performed on healthy staff or patients to establish their immune status or response to immunisation.

There are a variety of serological methods available to detect antibodies and these may measure total antibody or IgG or IgM separately. Guidelines for serological diagnosis are shown in Table B and further technical details of the methods used are described below.

As the presence of IgM antibody usually reflects acute infection, the availability of a test that detects IgM for the suspected organism often allows a single blood sample to be diagnostic if positive. In the absence of an IgM test, it is conventional to measure the 'titre' of antibody in two specimens: the 'acute' specimen is collected on presentation and the 'convalescent' specimen some 10 or more days later. Note that while the term convalescent is commonly used for the second specimen, the patient is not necessarily better and may have deteriorated! Paired specimens of this sort are classically tested by the complement fixation test (CFT).

Table B Three ways to make a serological diagnosis of recent or current infection.

- A positive IgM (or more rarely IgA) is (usually) diagnostic of recent or current infection with the organism tested (i.e. qualitative detection of specific IgM or IgA, usually by EIA or IF)
- IgG seroconversion (negative to positive in successive samples during the course of the current illness) is (usually) diagnostic of recent or current infection with the organism tested, usually by EIA or IF
- A rising titre (fourfold or greater increase) in successive samples is diagnostic of recent or current infection with the organism tested (e.g. quantitative detection of total antibodies by the complement fixation test)

Complement fixation test

The CFT measures the presence of (complement fixing) antibody which fixes complement only if it meets its target antigen. Thus the two serum samples (acute and convalescent) from the patient are mixed with a range of potential antigens appropriate for the clinical history (e.g. influenza and other respiratory organisms if a respiratory infection is suspected). A detection system (details of which are not important here) determines whether complement is used up (complement is said to be 'fixed') when the patient's serum is mixed with each selected antigen. If complement is fixed, this means antibody has bound to the antigen in question. In order to provide some quantitative measure of antibody level, a series of 'doubling dilutions' is prepared from each of the two serum samples (Fig. G). Since antibody production is a dynamic process during the course of an illness, the second (convalescent) sample would be expected to contain more antibody to the infecting organism than the acute sample (Table C). Antibodies to other organisms would, by contrast, remain more or less unchanged. By convention, a fourfold or greater increase

Table C Results of a complement fixation test following incubation of serial dilutions of both acute and convalescent sera with the suspected infecting organism in the presence of complement.

	Serum dilution					
	1 in 10	1 in 20	1 in 40	1 in 80	1 in 160	1 in 320
Acute serum	+	+	+	–	–	–
Convalescent serum	+	+	+	+	+	–

+ = antibody present. This is detected by the absence of complement, which has been used up ('fixed') by the antibody binding to its specific antigen. Each serum is eventually diluted to the extent that antibody is no longer detectable.

The fourfold rise in the titre of antibody from 40 to 160 during the course of this infection is deemed significant and is diagnostic of infection with the organism tested. A wide range of potential pathogens would normally be tested, but only the organism causing the current illness would reveal a rising titre of antibody.

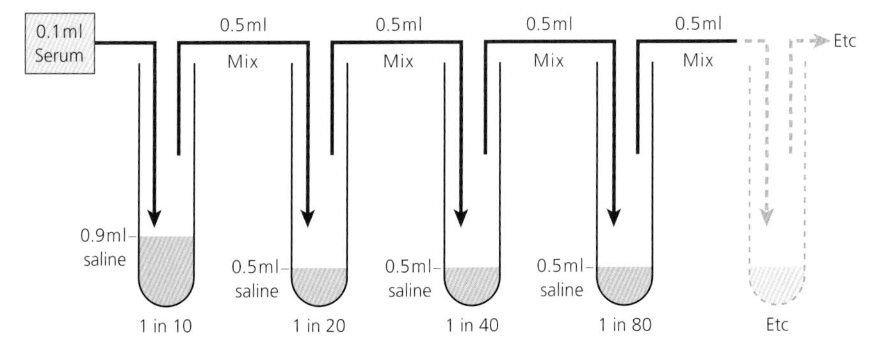

Figure G Preparation of doubling dilutions from a serum sample.

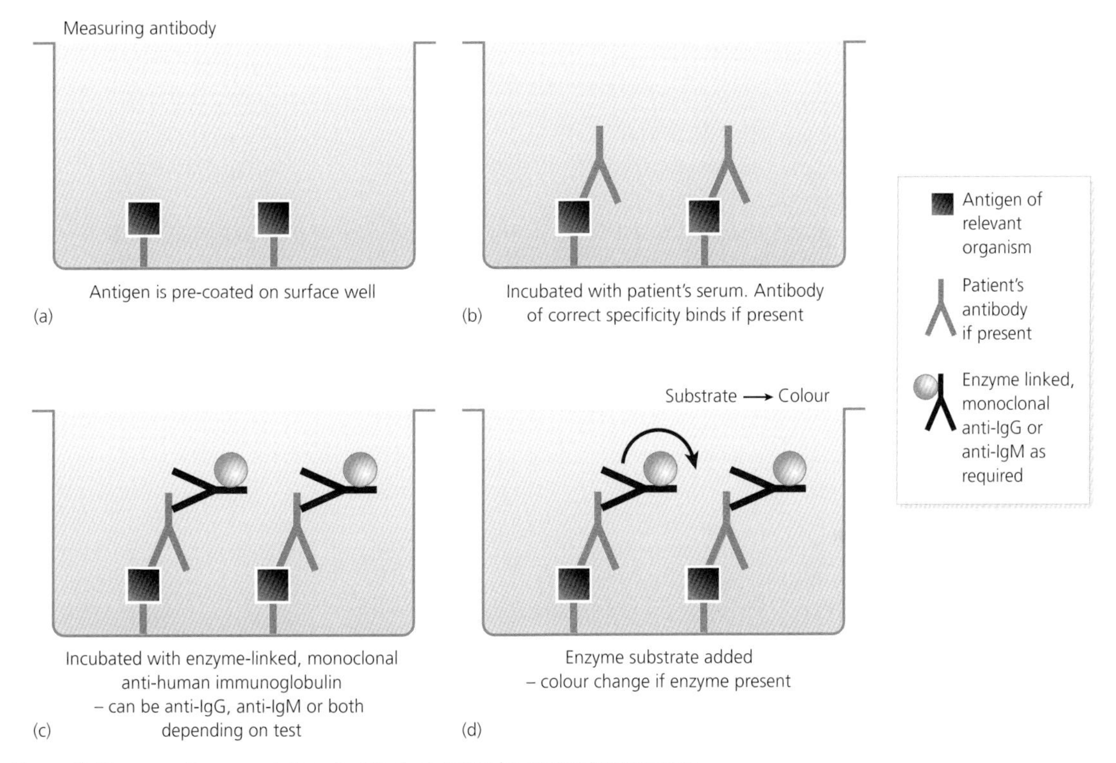

Figure H Diagrammatic representation of antibody detection by enzyme immunoassay.

in antibody titre (a 'rising titre') between the acute and convalescent samples is accepted as significant, i.e. diagnostic of recent infection. As the CFT detects antibodies indirectly by virtue of its ability to fix complement, it therefore does not distinguish between different antibody classes, but effectively measures a mixture of IgG and IgM.

Enzyme immunoassay, immunofluorescence and latex agglutination

Although it seems simple to experienced laboratory workers, one source of confusion for beginners is that the laboratory techniques of EIA, IF and LA can all detect either antigen or antibody, depending on how the test is designed. Remember that the key component in such

Table D Key points: approaches to the diagnosis of bacterial infection.

- M,C&S is the request accompanying most swabs, tissue or fluid specimens
- The clinical significance of bacterial growth must be interpreted with knowledge of expected normal flora and potential pathogens at the site of sampling. Not everything that grows will be reported
- Antimicrobial susceptibility tests are performed selectively on clinically relevant isolates with a limited range of appropriate antibiotics
- Antigen testing is useful for some bacterial infections
- NAAT is not yet widely used in routine bacteriology but this may change
- Serology is the most appropriate method of diagnosis for a few bacterial pathogens

Table E Key points: approaches to the diagnosis of viral infection.

- Microscopy (EM) and virus culture are now little used
- Antigen detection and NAAT have become the most important diagnostic methods for recent or current viral infections
- Use of real-time PCR is expanding and allows rapid diagnosis
- Serology is increasingly (but not exclusively) used to obtain evidence of previous infection or successful immunisation rather than for diagnosis of recent infection

tests is a monoclonal antibody specially selected for its specificity – this may be targeted at an antigen (e.g. hepatitis B surface antigen) or an antibody (e.g. IgM). A diagrammatic representation of an EIA designed to detect antibody is shown in Fig. H – compare this with the antigen detection EIA shown in Fig. E. Thus in interpreting the clinical significance of a result of such tests, it is important to check what is being measured.

Summary of laboratory diagnosis

The key points in the diagnosis of bacterial and viral infections are summarised in Tables D and E, respectively.

Antimicrobial chemotherapy

The term antimicrobial chemotherapy encompasses separate groups of agents that are active against bacteria (antibiotics), viruses (antivirals) and fungi (antifungals).

Antibiotics

Background

An antibiotic is said to be **bactericidal** if it kills organisms and **bacteriostatic** if it inhibits growth. The **minimum inhibitory concentration** (MIC) provides a quantitative measure of antibiotic activity against an organism. The MIC is the lowest concentration of that antibiotic at which growth of that particular organism is inhibited. Thus the lower the MIC, the more susceptible the organism is to that antibiotic. Pharmaceutical companies often quote MIC figures to demonstrate the effectiveness of their products and to compare them with competitor products. However, an antibiotic does not generally have the same MIC for all strains within a single species – there is usually a spectrum of activity. Thus in surveillance or research studies, a collection of strains might be tested – e.g. 100 isolates of *Escherichia coli* from urine – and the MIC_{50} and MIC_{90} determined for a relevant antibiotic. The $\mathbf{MIC_{50}}$ is the concentration of antibiotic that inhibits growth of 50% of strains tested, while the $\mathbf{MIC_{90}}$ is the concentration that inhibits 90% of strains tested. This gives a more meaningful overview of the likelihood that the antibiotic in question would be successful if used to treat an *E. coli* urinary tract infection. However, in order to interpret the clinical relevance of an MIC – that is, 'Does this MIC mean that an infection with this strain of this organism could be successfully treated with this antibiotic?' – you need further information. This requires knowledge of the concentration of the antibiotic that is clinically achievable at the site of infection, in this case in urine, without toxicity problems.

Infectious Disease: Clinical Cases Uncovered. By H. McKenzie, R. Laing, A. Mackenzie, P. Molyneaux and A. Bal. Published 2009 by Blackwell Publishing. ISBN 978-1-4051-6891-5.

There are a series of internationally agreed **breakpoints** that have resulted from consideration of such issues for all commonly used antibiotics and the breakpoint defines the MIC below which treatment is likely to be successful. In the treatment of serious infection, the MIC of the selected antibiotic for the infecting organism is a clinically useful piece of information and the tests required are easily performed in a routine diagnostic laboratory. MIC tests can be performed by preparing a series of dilutions of an antibiotic in liquid culture medium and then inoculating all dilutions with the organism to determine the level at which it will grow. This classical approach is still required from time to time, but more commonly an E-test is performed. This uses a paper strip accurately impregnated with a gradient of antibiotic concentrations – high levels at one end and low at the other. The pattern of growth of organism around this strip allows the MIC to be read off the antibiotic gradient scale (Fig. I).

Very occasionally, the **minimum bactericidal concentration** (MBC) is required. The MBC is the minimum concentration of the antibiotic required to kill the organism, not just inhibit its growth. The MBC for a particular combination of organism and antibiotic would normally be slightly higher than the MIC.

Susceptibility and resistance

Organisms isolated from clinical specimens are routinely reported as 'sensitive' (or susceptible) or 'resistant' to a range of antibiotics. An organism is considered **resistant** to a given drug when it is unlikely to respond to attainable levels of that drug in the tissues. It is useful to distinguish MIC testing (see above), which is a technical exercise producing a numerical result, from sensitivity testing, which is based on a laboratory test but also includes an interpretive step to decide on the basis of the result the likely outcome of treatment of a particular infection at a specified site with that antibiotic. A simple disc diffusion test can be used to identify resistant isolates

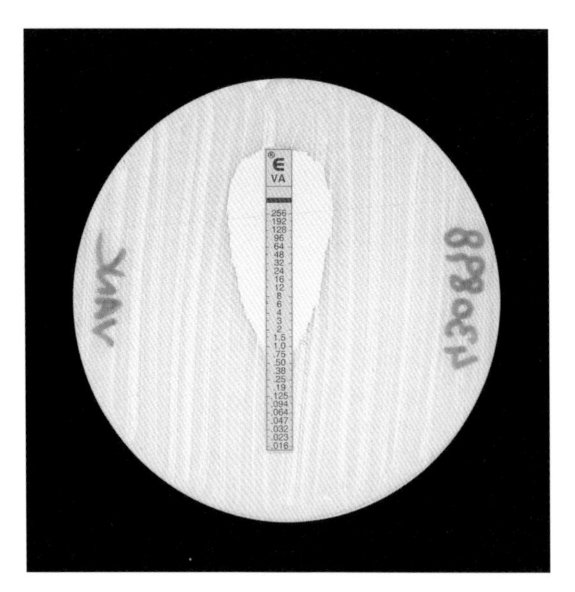

Figure I E-test method of minimum inhibitory concentration determination. The paper strip is calibrated with a range of antibiotic concentrations and the MIC can be read directly at the intersection of organism growth with the strip. In this example, the MIC of vancomycin for the staphylococcal strain tested is 0.75 mg/L.

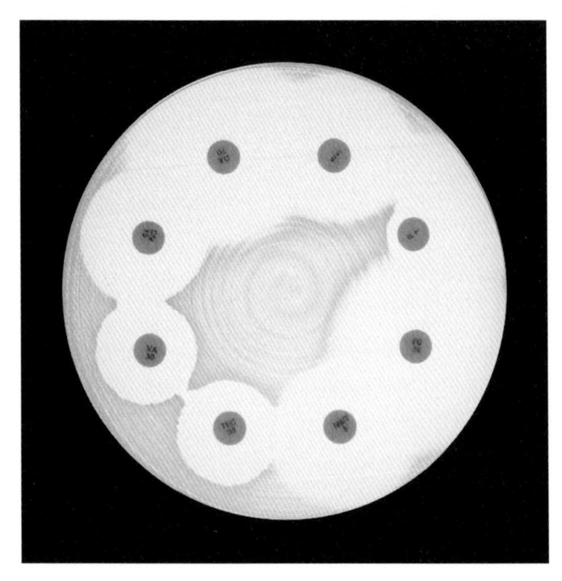

Figure J Disc diffusion test for antibiotic susceptibility. Published guidelines determine the size of the zone of inhibition around a disc that is required for the organism to be considered susceptible. In this example, the organism is resistant to penicillin (P).

(Fig. J), although a variety of semiautomated methods are in increasing use. Extensive quality control measures are required to ensure that such tests work reliably on a daily basis.

In some cases, all strains of a given species are intrinsically resistant to a drug and laboratory sensitivity testing is irrelevant, e.g. streptococci are always resistant to aminoglycosides and Gram-negative organisms are always resistant to vancomycin. The main focus of laboratory testing is therefore to detect acquired resistance among individual strains of a species for which the antibiotic in question might be an appropriate choice of treatment. Common mechanisms of resistance are described in Table F and clinically important antibiotic resistance mechanisms in Table G.

There are three genetic mechanisms by which resistance may arise:

1 *Chromosomal mutation.* Mutation can result in the development of resistant forms of bacterial species which will be selected or encouraged by the presence of antibiotic.
2 *Chromosomal exchange.* Resistance genes are often spread between organisms on mobile elements of DNA called **transposons**, which are able to incorporate themselves into the chromosome of their new host.
3 *Plasmid mediated.* Plasmids are packages of extrachromosomal DNA that frequently carry resistance genes. The ability of plasmids to spread readily both within and between bacterial species means that resistance can increase rapidly if the antibiotic in question is widely used.

Table F Mechanisms of antibiotic resistance.

Mechanism	Example
Organism produces an enzyme that breaks down the antibiotic	β-lactamase enzymes attack penicillins and cephalosporins
Genetic alteration in the antibiotic target site within the organism renders the antibiotic ineffective	Methicillin-resistant *Staphylococcus aureus* (MRSA) strains have an altered target site to which β-lactam agents cannot bind
Organism acquires an efflux pump that pumps the antibiotic out of the cell	Some forms of macrolide resistance result from a macrolide-specific efflux pump

Table G Summary of clinically important antibiotic resistance mechanisms.

Organism	Mechanism of resistance	Comment
Staphylococcus aureus	β-lactamase produced by the organism cleaves the β-lactam ring and thus deactivates the antibiotic	Infection can still be treated with β-lactamase stable antibiotic (flucloxacillin) or by adding a β-lactamase inhibitor (co-amoxiclav, piperacillin-tazobactam)
Methicillin-resistant *Staphylococcus aureus* (MRSA)	The organism has acquired a series of resistance genes that result in an altered penicillin-binding protein target	Penicillins or cephalosporins have no activity as their target site has altered. Can be treated with glycopeptides or newer agents such as linezolid or daptomycin
Glycopeptide intermediate *Staphylococcus aureus* (GISA)	Thickened cell wall with overproduction of target site	These infections are fortunately rare
Streptococcus pneumoniae	The organism has acquired a series of resistance genes that result in an altered penicillin-binding protein target	Intermediate and high level penicillin resistance in pneumococci is a clinical problem in some countries. Unlike MRSA, it is still possible to treat some infections with high-dose ceftriaxone if the MIC is not too high and reasonable levels can be achieved at the site of infection
Enterococci	Vancomycin-resistant enterococci (VRE) contain the van (A) or van (B) gene which changes the structure of the cell wall target	Can be very difficult to treat. Expert opinion should be sought
Enterobacteriaceae	Extended spectrum β-lactamase (ESBL) renders organisms resistant to all penicillins and cephalosporins	Serious infections should be treated with carbapenems (e.g. imipenem or meropenem)

Major groups of antibiotics

Table H summarises the antibiotics commonly used in UK hospitals.

Beta lactams

Beta lactams inhibit cell wall synthesis by inhibiting the enzymes that cross link peptidoglycan in the bacterial cell wall. The enzymes that are inhibited by β-lactams have become known as **penicillin-binding proteins** (PBPs), although of course this is not their primary function as far as the bacterium is concerned. Once cell wall synthesis has been disrupted, the organism is killed by autolytic enzymes. Since human cells do not have a cell wall, they are unaffected by these antibiotics and as a result these agents are relatively safe. Bacteria have evolved two major resistance mechanisms to β-lactams:

1 The production of the enzyme β-lactamase destroys the activity of the antibiotic by breaking down the structurally important β-lactam ring.

2 Alterations in the structure of the bacterial PBPs can prevent β-lactams from binding to them.

The two main subgroups of β-lactams are penicillins and cephalosporins.

Penicillins

- **Benzylpenicillin** (penicillin G) was the original naturally occurring β-lactam first discovered by Sir Alexander Fleming. It still remains a very useful drug for certain indications, but many synthetic derivatives have been produced with an extended spectrum of action. In particular, many Gram-negatives are resistant to benzylpenicillin because of the relative impermeability to the cell wall.
- **Amoxicillin** has better oral absorption than benzylpenicillin and initially it had better Gram-negative activity, but nowadays coliform organisms are often resistant due to β-lactamase activity.
- **Co-amoxiclav** combines amoxicillin with the β-lactamase inhibitor clavulanic acid, thus extending

Table H Summary of antibiotics in common use in hospitals.

Antibiotic	Common usage	Comments	Antibiotic class
Benzylpenicillin	*Neisseria meningitidis, Streptococcus pneumoniae* and *S. pyogenes*	Preferred treatment for these infections once diagnosed (except penicillin-resistant pneumococci); effective and narrow spectrum. However, always be aware of penicillin allergy	β-lactam (penicillin)
Flucloxacillin	*Staphylococcus aureus*	Resistant to actions of β-lactamase, but no action against MRSA	β-lactam (penicillin)
Co-amoxiclav	Gram-negatives when susceptible (not *Pseudomonas*) and streptococci, enterococci and *S. aureus* (not MRSA). Also has good anaerobic activity	Includes the β-lactamase inhibitor clavulanic acid	β-lactam (penicillin)
Piperacillin/ tazobactam	Gram-negatives including *Pseudomonas*. Also active against enterococci and anaerobes	Tazobactam is a β-lactamase inhibitor. Pip/taz is often used in combination with gentamicin for serious intra-abdominal infection, or for empirical treatment of neutropenic sepsis	β-lactam (penicillin)
Ceftriaxone	Gram-negatives (not *Pseudomonas*) and streptococci. Widely used for pneumonia and meningitis – good activity against pneumococci	Does not act on enterococci	β-lactam (cephalosporin)
Clarithromycin	'Atypical pneumonia', staphylococci and streptococci when susceptible	Often used for Gram-positive infection if patient is allergic to penicillin	Macrolide
Clindamycin	Staphylococci and streptococci when susceptible; good anaerobic cover	Good tissue penetration so commonly used for cellulitis, necrotising fasciitis and bone and joint infections. *Clostridium difficile* associated diarrhoea is a well recognised side effect	Lincosamide
Ciprofloxacin	Gram-negatives including *Pseudomonas*	Excellent oral bioavailability and often the only oral agent available for *Pseudomonas* infection	Fluoroquinolone
Gentamicin	Gram-negatives including *Pseudomonas, S. aureus* but not streptococci	Intravenous only; monitor levels due to toxicity. Once-daily dosing now popular	Aminoglycoside
Vancomycin	Acts only on Gram-positive organisms; major use for serious MRSA infection	Nephrotoxic so monitor levels. Intravenous use only, with the exception of oral use for *C. difficile* infection	Glycopeptide
Trimethoprim	Urinary tract infection		
Meropenem	Wide Gram-positive and Gram-negative spectrum including anaerobes and extended spectrum β-lactamase (ESBL) producers	Not active against enterococci	Carbapenem
Metronidazole	Anaerobes	Also used as an anti-parasitic agent	

its spectrum to cover many β-lactamase-producing coliforms.

• **Flucloxacillin** is resistant to the actions of staphylococcal β-lactamase and has therefore been until recently the first choice for the treatment of staphylococcal infection. However, it is not active against MRSA (Table H). Methicillin is similar to flucloxacillin and is used to represent flucloxacillin in laboratory testing, hence the label methicillin-resistant *Staphylococcus aureus* (MRSA).

• **Piperacilllin** is a broad spectrum penicillin which is only used intravenously. It has useful activity against *Enterococcus* species and is active against *Pseudomonas*. More recently, its spectrum has been extended by use in combination with the β-lactamase inhibitor tazobactam. Piperacillin-tazobactam also has good anaerobic activity.

• **Carbapenems** are a new subgroup of penicillins with a very wide spectrum of activity, including anaerobes. Those in current use are imipenem, meropenem and ertapenem.

Cephalosporins

This is a large and confusing group of antibiotics with a wide range of uses. They are typically grouped in four 'generations', more or less in the chronological order that the cephalosporins were introduced. Among cephalosporins, activity against Gram-negative organisms increases from first generation drugs (e.g. cephradine) through second generation (e.g. cefuroxime), third generation (e.g. ceftriaxone) and fourth generation (e.g. cefepime). Amongst third generation cephalosporins, only ceftazidime is active against *Pseudomonas* species. In contrast, Gram-positive activity (e.g. against staphylococci) decreases from first through to third generation drugs. Fourth generation cephalosporins, on the other hand, have a wide spectrum of activity. They are active against Gram-positive organisms such as methicillin-susceptible *S. aureus* and streptococci and also against Gram-negative bacilli including both Enterobacteriaceae (e.g. *E. coli, Klebsiella, Enterobacter*) and *Pseudomonas*.

• Oral cephalosporins (mostly first generation, e.g. **cephalexin**) are used widely in general practice to treat urinary and respiratory tract infections.

• **Cefuroxime** is used intravenously in hospitals, often as a prophylactic antibiotic. As a second generation cephalosporin, it has a reasonable spectrum of both Gram-positive and -negative activity.

• **Ceftriaxone** (cefotaxime is very similar apart from the dosing regimen) has better MICs for Gram-negative organisms than cefuroxime and also gives good cover against streptococci, but not enterococci. Ceftriaxone is therefore widely used for chest infections (covers pneumococci and *Haemophilus influenzae*), meningitis (pneumococci and meningococci) and along with metronidazole for surgical prophylaxis. However, its wide spectrum can be a disadvantage and widespread usage is linked to high rates of *Clostridium difficile* infection.

Glycopeptides

The glycopepetides, **vancomycin** and **teicoplanin**, act on a different part of the cell wall synthesis pathway to β-lactams and have activity only against Gram-positive organisms. Both are only used intravenously (with one major exception – oral vancomycin is used for *C. difficile* infection); vancomycin levels must be monitored to reduce potential nephrotoxicity. Newer glycopeptides that have longer dosing intervals are being developed.

Aminoglycosides

Aminoglycosides act on protein synthesis and are noted for their activity against Gram-negative organisms including *Pseudomonas*, with low levels of resistance seen in the UK. They have good staphylococcal activity as well, but streptococci are intrinsically resistant. **Gentamicin** is the most commonly used aminoglycoside, but serum levels must be monitored because of nephrotoxicity and ototoxicity. Once-daily dosing has been shown to maintain efficacy but to reduce side effects, so this is now commonly used. A high single dose based on body weight is given and levels checked between 6 and 14 hours later. The dosing interval can then be determined by interpolation of the level on a dosing chart.

Other aminoglycosides are used occasionally as there can be differences in resistance to individual agents within this group. Thus tobramycin and amikacin are sometimes used in special circumstances (e.g. in cystic fibrosis) when Gram-negative isolates are resistant to gentamicin.

Macrolides, lincosamides and streptogramins

These three groups of antibiotics are not structurally related, but because they all inhibit protein synthesis by acting at the ribosome, there is often, but not always, cross resistance between them and they tend to be grouped together.

• The macrolides **clarithromycin** and **erythromycin** are active against Gram-positive organisms and are often used in patients with penicillin hypersensitivity.

Macrolides are effective against the organisms that cause 'atypical pneumonia', e.g. *Chlamydia psittaci, Coxiella burnetii* and *Mycoplasma* pneumoniae, and are first choice against *Legionella pneumophila.* The newer macrolide, **azithromycin**, is useful for single-dose treatment of genital infection with *Chlamydia trachomatis.*

• **Clindamycin** is a lincosamide and it is occasionally useful to know that it has a different structure from macrolides and can still be used in patients who have had allergic reactions to the latter. It has a similar spectrum of activity as macrolides against Gram-positive bacteria but also has useful anaerobic activity. A particular strength of clindamycin is its good oral absorption and good penetration into the tissues – the latter means it is particularly good for bone and joint infections and for cellulitis or necrotising fasciitis. The well known disadvantage is the association between clindamycin and pseudomembranous colitis. However, it must be appreciated that all antibiotics promote *Clostridium difficile* infection and the risk of using clindamycin must be weighed against the need to give microbiologically optimised treatment in serious infection.

• Streptogramins are not widely used, but the combination of quinupristin and dalfopristin (QD) is available. QD is active against MRSA but is only available for intravenous use. In practice, QD is used only when other agents are contraindicated.

There are two major patterns of resistance seen in streptococci and staphylococci to these three groupings of antibiotics. M phenotype resistance involves resistance to macrolides only and is mediated by an efflux pump. Thus clindamycin is still active against such organisms. In MLS_B resistance, the organism modifies its ribosomal target site so that none of these antibiotic groups have any activity. Both M and MLS_B resistance patterns are frequently linked with tetracycline resistance, not because the mechanisms of resistance are shared, but because the resistance genes frequently travel together on the same transposon.

Tetracyclines

Tetracyclines inhibit bacterial protein synthesis by binding to the 30S ribosomal subunit. **Oxytetracycline, doxycycline** and **minocycline** are within this group. Tetracyclines are active against *Chlamydia, Rickettsia, Brucella* and *Borrelia* (the causative agent of Lyme disease). They are also active against many other groups of bacteria but are rarely used as first-line antibiotics. Notably, doxycycline is useful for malaria prophylaxis.

Tetracyclines should not be prescribed to pregnant women, breast-feeding women and children under 12 years of age. This is because tetracyclines affect bone development.

Bacteria may acquire resistance to tetracyclines by efflux pumps or on account of 'ribosomal protection proteins' that prevent the binding of tetracyclines to the ribosomes.

Tigecycline is a newer, long acting tetracycline that has a broad spectrum of activity. It also binds more strongly to ribosomes than older tetracyclines and is a poor substrate for efflux pumps. Thus, tetracycline-resistant strains can still be susceptible to tigecycline. Tigecycline is active against MRSA and against Gram-negative bacteria that produce extended spectrum β-lactamases. Tigecycline is not active against *Pseudomonas.*

Fluoroquinolones

The quinolones act on DNA gyrase, an enzyme involved in maintaining the superfolded structure of DNA. **Ciprofloxacin** is the most commonly used and has good activity against Gram-negative organisms including *Pseudomonas.* Ciprofloxacin is well absorbed orally and therefore provides virtually the only option for oral therapy in the treatment of *Pseudomonas* infections. Ciprofloxacin has some activity against staphylococci but would not be a first choice agent for this purpose. Ciprofloxacin does not generally have good activity against streptococci. However, **levofloxacin** is a newer fluoroquinolone with good activity against the pneumococcus and is also effective against the organisms causing 'atypical pneumonia', so it is a possible choice in the treatment of community-acquired pneumonia.

Urinary tract agents

Trimethoprim and **nitrofurantoin** are commonly used antibiotics for urinary tract infections. Trimethoprim inhibits the enzyme dihydrofolate reductase enzyme in bacteria, while sulfonamides inhibit dihdyropteroate synthetase in the same series of reactions that are involved in folate synthesis. Thus, the combination of trimethoprim and sulfamethoxazole, commonly known as co-trimoxazole, has useful synergistic activity. However, this combination is not commonly used because of toxicity problems. Trimethoprim is well absorbed and is excreted in the urine where it exerts its antibacterial activity.

Nitrofurantoin has a wide spectrum of activity encompassing both Gram-positive and Gram-negative bacteria. *Pseudomonas* and *Proteus* are resistant to this agent but

E. coli, *Citrobacter*, streptococci and enterococci are usually susceptible. Nitrofurantoin has no role in the treatment of systemic infection and is used only in urinary tract infection.

Anaerobic agents

Metronidazole is used in the treatment of anaerobic infections. It also has anti-protozoal activity and is used to treat *Trichomonas vaginalis* infection in the genital tract.

New drugs in novel antibiotic classes

• **Linezolid** is an oxazolidinone. This new drug is active only against Gram-positive bacteria and the most common indication is infection caused by MRSA. Linezolid has excellent oral bioavailability and tissue penetration. However, it can lead to serious adverse effects such as haematological toxicity and neuropathy and should not be used for treatment beyond 28 days. Widespread resistance has not been documented so far.

• **Daptomycin** is a lipopeptide antibiotic that causes cell membrane depolarisation in Gram-positive bacteria and resultant efflux of potassium ions. It is available for intravenous use only and is licensed for skin and soft tissue infections caused by Gram-positive bacteria. It has also been found to be useful in right-sided staphylococcal endocarditis. Patients receiving daptomycin should have their creatine kinase levels monitored because muscle toxicity is a potential side effect.

Antiviral drugs

Specific drugs are required for the treatment of viral infections as antibiotics have no action against these microorganisms. There are no virucidal agents and they are all virustatic, i.e. they inhibit growth or replication of the virus but do not kill it. Only a few viral diseases are treatable and they tend to be serious and potentially fatal conditions rather than those that are common and self-limiting.

Anti-herpes virus drugs

Herpes viruses are a clinically important group of viruses, all of which become latent. They include herpes simplex virus (HSV), Epstein–Barr virus (EBV), varicella zoster virus (VZV) and cytomegalovirus (CMV). These are not equally sensitive to antiviral agents but treatment is most effective if started early in the infection. The drugs are largely nucleoside analogues that interfere with DNA nucleic acid synthesis.

• **Aciclovir** is a nucleoside analogue that is converted into its active form by an enzyme (thymidine kinase) which is coded for by the viral genome. Thus it is specific for viral infected cells and has very low toxicity for uninfected host cells. Aciclovir is extremely active against HSV and is active against VZV. The intravenous form is used to treat serious infection such as herpes encephalitis and VZV pneumonitis. Cold sores, caused by HSV reactivation, can be treated topically if given when prodromal symptoms start and before ulceration occurs.

• **Famciclovir** and **valaciclovir** are oral agents related to aciclovir but with better bioavailability. They are used for treatment of HSV and shingles.

• **Ganciclovir**, also a nucleoside analogue, is used intravenously against CMV. However, it is toxic and its use is limited to life-threatening or sight-threatening infections in the immunocompomised (e.g. acquired immune deficiency syndrome (AIDS), transplant recipients). **Valganciclovir**, a pro-drug of ganciclovir, is an oral alternative to ganciclovir for some CMV infections, both for treatment and prophylaxis. Bone marrow toxicity means that blood must be monitored carefully for both drugs.

• **Foscarnet** is a different category of drug that can be used for some HSV, VZV and CMV infections which are resistant to nucleoside analogues. It is highly nephrotoxic and can only be given intravenously.

• **Cidofovir** is used for CMV retinitis when other antiviral drugs are inappropriate.

Anti-HIV drugs

Ziduvudine (AZT), a nucleoside analogue, was the first anti-human immunodeficiency virus (HIV) agent and inhibits reverse transcriptase, an essential enzyme in the process of HIV replication. However, it has major side effects including anaemia and neutropenia, and the virus rapidly developed resistance when AZT was used alone. Combination therapy (using a combination of more than one type of drug) quickly became established as the most effective way of prolonging survival in AIDS patients and is now standard practice. Patients usually receive two **nucleoside analogue reverse transcriptase inhibitors** in combination with either a **non-nucleoside analogue reverse transcriptase inhibitor** (e.g. efavirenz) or a **protease inhibitor** (e.g. saquinavir) which inhibits viral protease enzyme. Choice of therapy and response to treatment is assessed by monitoring the patient's HIV viral load and CD4 count (see Infection and immunity, Part 1).

Similar combinations of drugs are used for prophylaxis following occupational or sexual exposure to HIV-positive blood or body fluids and to decrease the risk of transmission from HIV-positive mothers to the unborn or newborn child. All anti-retroviral drugs have significant toxicity, including the long-term effects of raised lipid levels that are caused by some drugs. The recommended treatment of HIV changes as new therapies become available. For further information, including information on drug classes not mentioned here, consult a suitable website such as www.bhiva.org.

Drugs for chronic hepatitis B and C

Interferon-α is a protein produced as part of the normal human immune response. A genetically engineered form of this is used to treat chronic hepatitis B and C infections, although the low response rate, serious side effects and high cost have limited its use. The genetically engineered product is linked to polyethylene glycol (pegylated) to slow excretion and thus reduce treatment frequency to once a week. Combination therapy with weekly pegylated interferon-α given by subcutaneous injection and daily oral ribavarin (see below) is currently the best treatment for chronic hepatitis C infection, but this does not suit all patients. Other patients may receive tablet therapy with drugs such as **lamivudine**, a nucleoside analogue that is also used in HIV treatment. Other drugs that are also of use in chronic hepatitis B treatment include **adefovir, tenofovir, entecavir** and **telbivudine**.

Drugs for viral respiratory infections

Zanamavir and **oseltamivir** are both licensed for the treatment of influenza A and B within 48 hours of the onset of symptoms. Oseltamivir is also licensed for post exposure prophylaxis. There is a limited role for these drugs in seasonal influenza (see the *British National Formulary*) and immunisation remains the first line of protection. The UK Health Departments have stores of oseltamivir as part of contingency planning for pandemic influenza.

Ribavarin is another nucleoside analogue which is used for the treatment of severe respiratory syncytial virus (RSV) infection. However, as it must be inhaled as a fine spray to reach the site of infection in the lungs and administration in young children is difficult, it is little used for this. Ribavarin has also been shown to reduce mortality in Lassa fever and is used in combination treatment for hepatitis C (see above).

Antiviral resistance

Testing viruses for resistance is a relatively new science compared to antibacterial resistance testing, but there are a number of situations in which genotyping, by sequencing parts of the viral genome, helps choose the most effective treatment strategy. Testing for drug resistance is mainly used for HIV patients, looking for specific mutations in the HIV genes that are known to be linked to resistance to the different HIV drugs.

Antifungal drugs

The majority of antibiotics have no activity against fungi and fungal infections must be treated with a different range of drugs. Fungi may be subdivided into yeasts and filamentous fungi, although a few can take either form (dimorphic fungi). Antifungal agents differ in the extent to which they cover these various forms. There are several classes of antifungal agents.

Polyenes

Polyene drugs bind to ergosterol, which is a component of the fungal cell wall but not the bacterial cell wall. This results in increased permeability of the cell wall and these drugs are active against both yeasts and filamentous forms. Unfortunately, polyenes also bind to other sterols (e.g. cholesterol) in mammalian cells and this is the reason for their toxicity.

• **Amphotericin B** is the only polyene available for intravenous use and is used for the treatment of serious systemic fungal infection. Its side effects include renal, hepatic and cardiac toxicity. A lipid formulation of the drug – liposomal amphotericin – offers reduced incidence of such side effects, but amphotericin B should only be used when clinically indicated for serious systemic fungal infection. Resistance is unusual and conventional sensitivity testing is not routinely performed as the *in vitro* results do not appear to correlate well with clinical outcome.

• **Nystatin** is the other polyene drug in common clinical use. Unlike amphotericin B, it is available for topical use only (e.g. creams for fungal skin infection, pessaries for vaginal *Candida* infection and an oral suspension for oral and oesophageal candidiasis).

Azoles

These drugs inhibit ergosterol synthesis and those in common clinical usage include **fluconazole, itraconazole** and **voriconazole**. Fluconazole has been used widely for oral and parenteral treatment of yeast

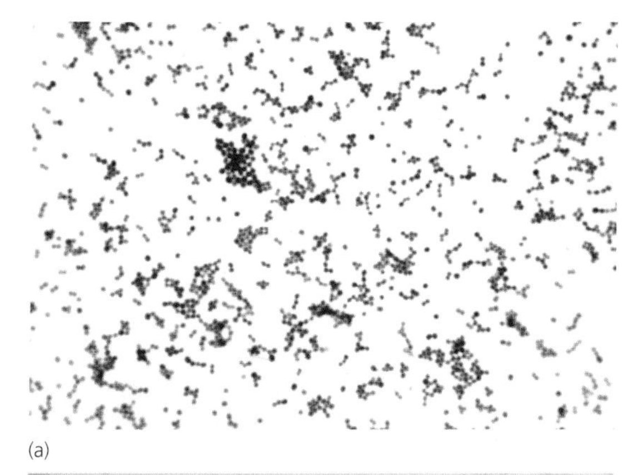

(a)

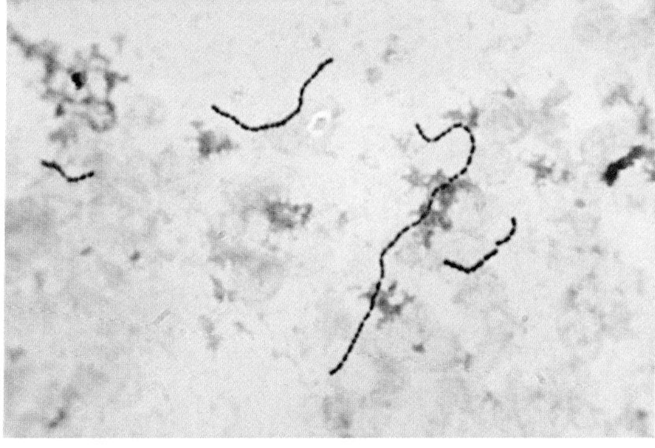

(b)

Plate 1 Example Gram stain showing Gram-positive cocci: (a) in clusters (staphylococci) and (b) in chains (streptococci).

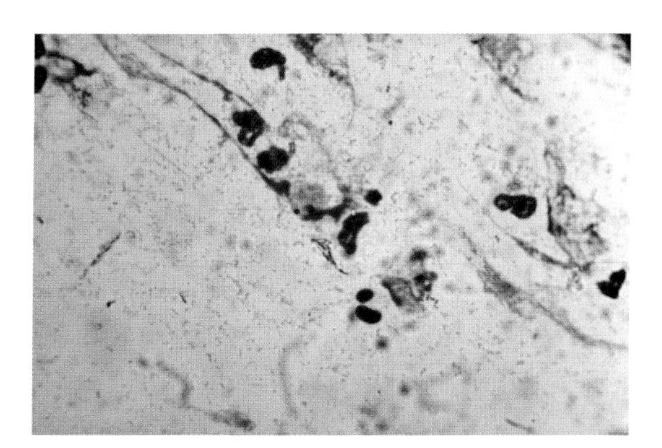

Plate 2 Example Ziehl–Neelsen stain showing red alcohol and acid fast bacilli (AAFB) against a blue counterstain.

Plate 3 Bacterial culture of a wound swab on blood agar. The three stages shown in the pictures are: (a) inoculation of the swab over one-third of the surface of the agar; (b) spreading of any bacteria present across the surface with a sterile plastic disposable loop; and (c) the agar plate after overnight culture showing individual bacterial colonies.

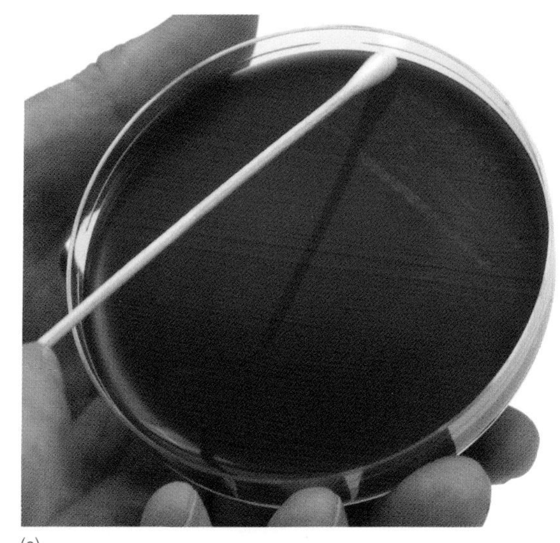

(a)

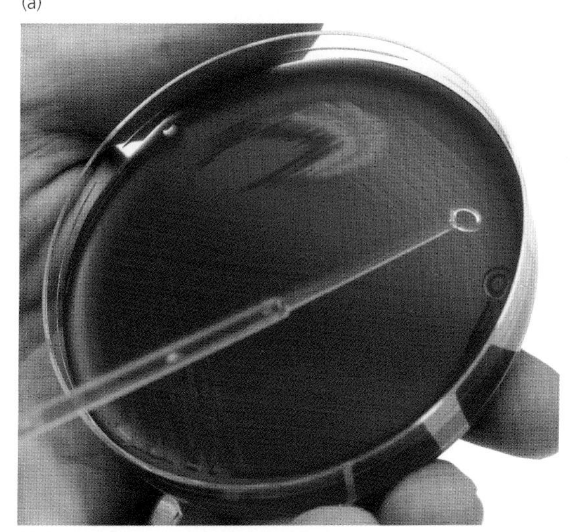

(b)

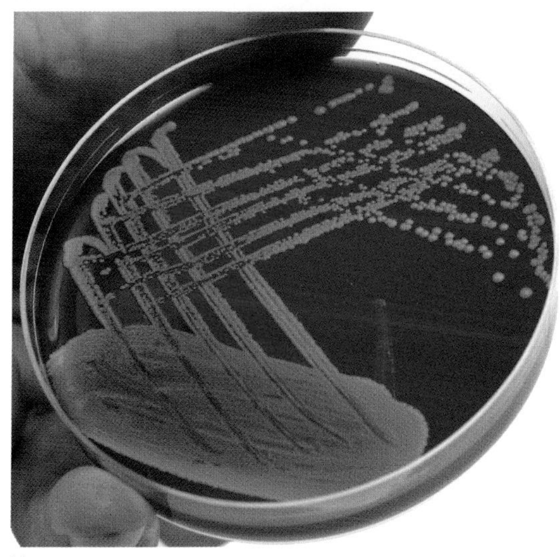

(c)

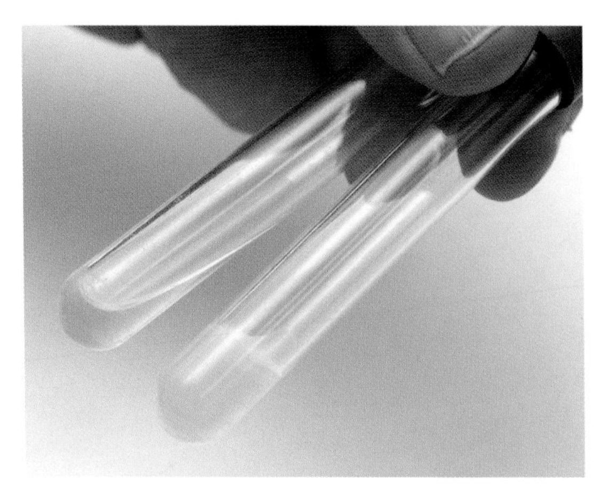

Plate 4 Coagulase test showing the ability of *Staphylococcus aureus* to coagulate plasma (tube on right) while a wide range of other staphylococcal species (coagulase-negative staphylococci) cannot (tube on left).

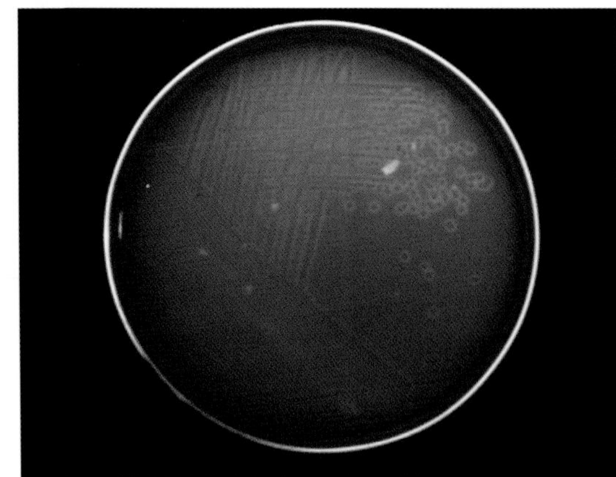

(a)

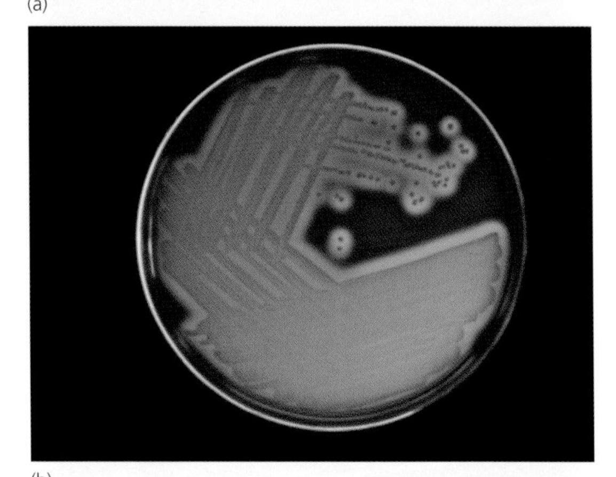

(b)

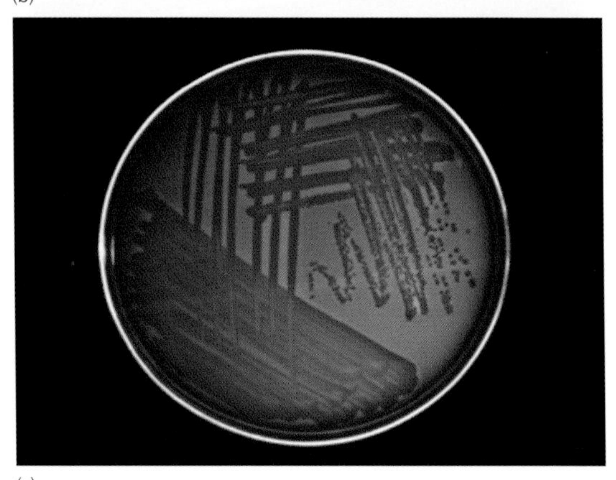

(c)

Plate 5 Haemolysis of blood agar by streptococci showing: (a) α-haemolysis (partial), (b) β-haemolysis (complete) and (c) no haemolysis.

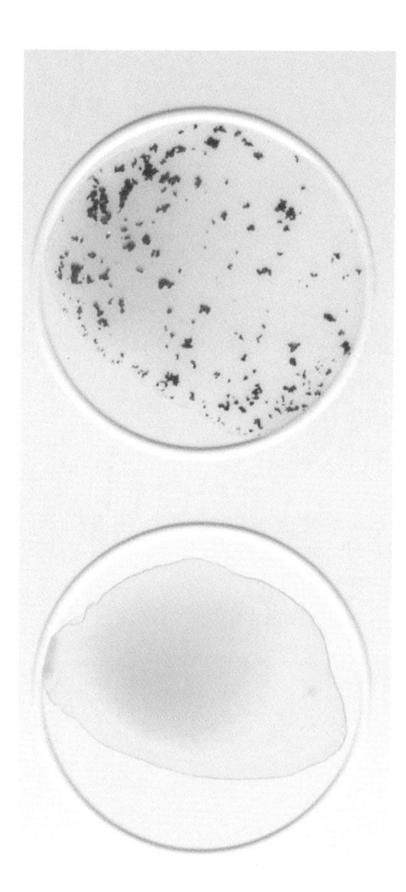

Plate 6 Example of a latex agglutination test showing positive (top) and negative (bottom) reactions.

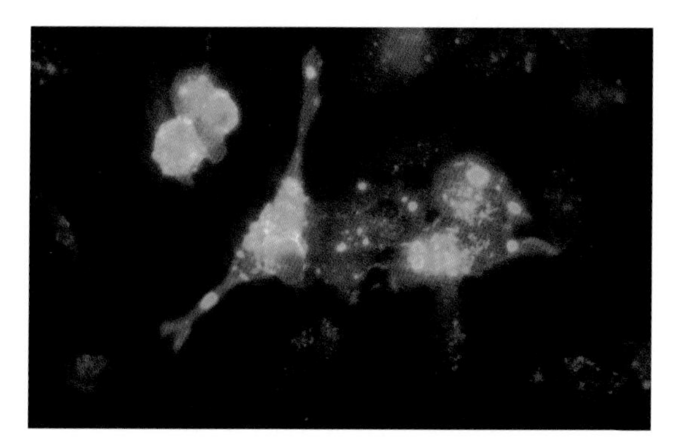

Plate 7 Example of direct immunofluorescence demonstrating the presence of respiratory syncytial virus (light green staining) in a nasopharyngeal aspirate from a child. Courtesy of Mr Ian Collacott.

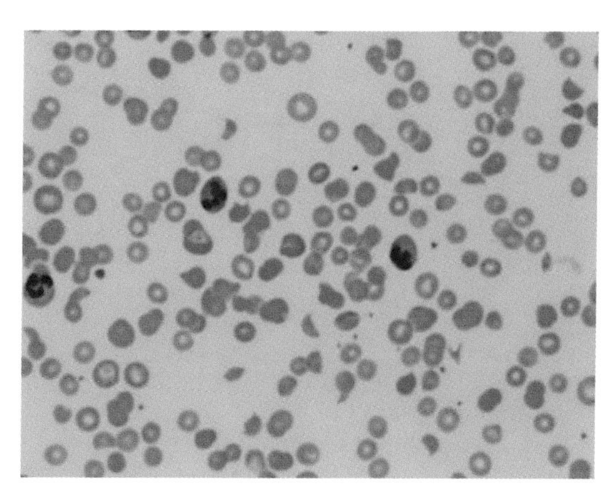

Plate 8 Blood film showing severe microangiopathic haemolysis with schistocytes (fractured red blood cells) in a patient with haemolytic uraemic syndrome. Courtesy of Dr Ghada Zakout.

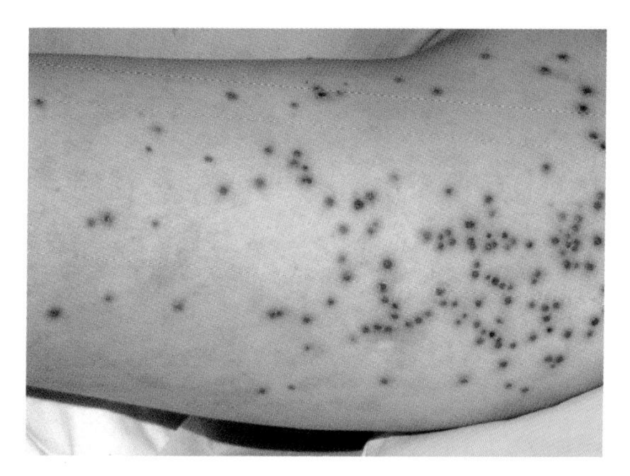

Plate 9 Typical vesicular and pustular lesions in a patient with chickenpox.

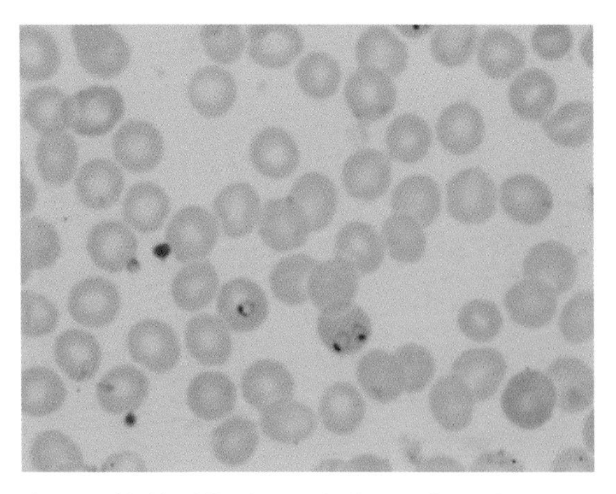

Plate 10 Thin blood film showing the fine ring forms of *Plasmodium falciparum* inside the red blood cells. Courtesy of Dr Ghada Zakout.

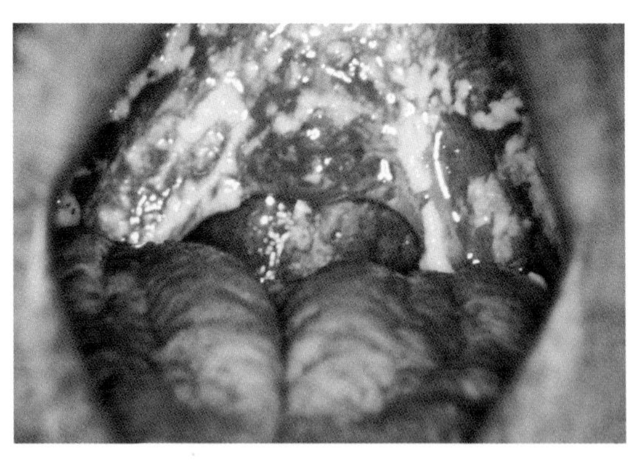

Plate 12 An example of extensive oral candidiasis. Courtesy of Dr C. C. Smith.

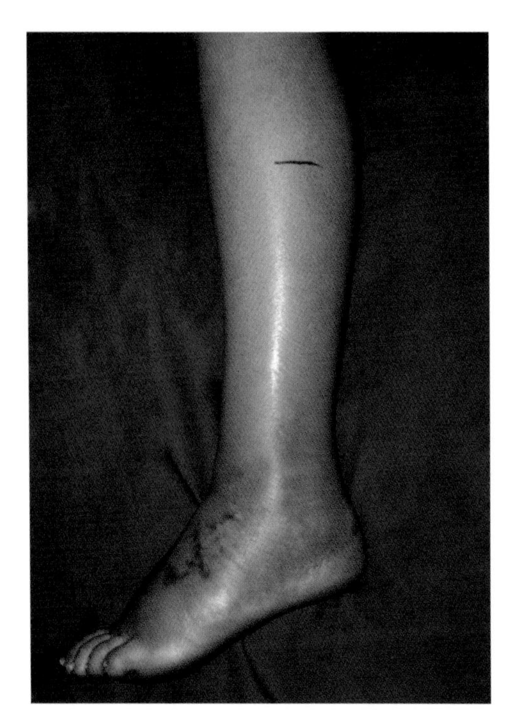

Plate 11 Cellulitis of the left leg. The pen mark indicates the upper extent of erythema at the time of admission to hospital. Courtesy of Dr C. C. Smith.

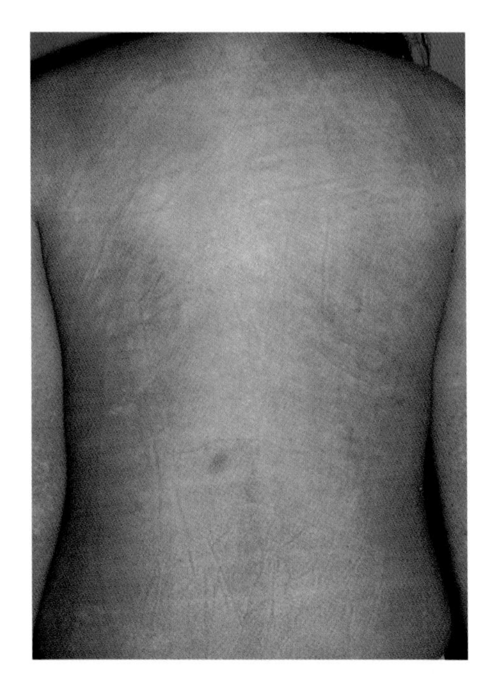

Plate 13 Diffuse macular rash.

infections, but has no activity against filamentous fungi such as *Aspergillus* spp. and not all yeasts are sensitive. *Candida albicans*, which is the commonest clinical isolate, is usually sensitive, but species such as *C. krusei* are resistant. Itraconazole is active against both yeasts and filamentous fungi, including *Aspergillus* spp. and dermatophytes. Voriconazole is used to treat invasive aspergillosis but can cause serious adverse effects and drug interactions.

Allylamines

Allylamines also suppress ergosterol synthesis but act at a different stage of the pathway from azoles. The only allylamine in common use is **terbinafine**, which is active against dermatophytes. Clinical use is therefore restricted to dermatophyte infection of the skin (e.g. ringworm, athlete's foot) and nails (onchomycosis). Mild infections are treated topically and more serious infections (including onchomycosis) are treated orally.

Echinocandins

Echinocandins inhibit the synthesis of glucan polysaccharide in several types of fungi. Caspofungin is an echinocandin in common use and is only available intravenously. It is used for the treatment of serious *Candida* and *Aspergillus* infections and is used as an empirical antifungal agent in patients with neutropenic sepsis.

Infection and immunity

Infection is surprisingly difficult to define. The presence of a microorganism in or on the human body can give rise to the signs and symptoms of infection, but can also represent colonisation or commensalism, in which the microorganism is part of the normal flora and does not result in symptomatology or pathology of any kind. The outcome of the interaction between a microorganism and the human host depends upon the pathogenicity of the particular strain of organism and the effectiveness of the individual human immune system in combating it. There are several possible outcomes when a human host and a microorganism interact for the first time:

- The organism is immediately removed/destroyed.
- The organism establishes itself in/on the host but does not result in clinical symptoms (colonisation).
- The organism establishes itself in/on the host and causes signs and/or symptoms of disease.

Thus an understanding of the pathophysiology of infection depends upon understanding the interaction between the microorganism and host immunity.

Organism characteristics

The pathogenicity of an organism is the capacity of that organism to cause disease. This is determined by two features, infectivity and virulence:

- Infectivity is the ability of an organism to become established on or within a host.
- Virulence is the ability of the organism to cause harmful effects once established.

Thus rhinoviruses, which cause the common cold, are of high infectivity but low virulence, while avian influenza is of low infectivity for humans (at least at present) but is highly virulent as it has a high fatality rate if it does become established in the human host.

- **Infectivity** is determined by factors such as the ability to attach to mucosal surfaces and survive in the host environment.
- **Virulence** is determined by factors such as the production of toxins (at least in bacteria) which cause cell damage or enzymes which promote invasion through the tissues. There are a variety of toxin types:
  - **Exotoxins**: toxins that are excreted by the cell and can have their pathological effects at some distance from the site of infection (including enterotoxins that act on the gastrointestinal tract).
  - **Endotoxin**: a constituent part of the Gram-negative cell wall with an important role in septic shock.
  - **Superantigen**: exotoxins of *Staphylococcus aureus* and *Streptococcus pyogenes* that stimulate T cells (see below).

Host immunity

The human immune response is complex and much of the detail is outwith the scope of this book. However, it is not possible to understand infection if we consider only the causative microorganism. An overview of the basic components of immunity is given below.

Innate immunity

Innate immunity encompasses protective mechanisms that do not differentiate between different pathogens and do not have the capacity to create immunological memory, i.e. it is not a 'learned' response. However, the effectiveness of such processes only becomes evident when they are impaired in some way. Simple examples of the importance of innate immunity are listed in Table I.

Phagocytosis

The phagocytic cells – macrophages and (neutrophil) polymorphs – are part of the innate system and ingest (phagocytose) microorganisms without any specific recognition process. Polymorphs are a major component of the acute immune response and have a short life, while macrophages are involved in chronic inflammation and

Infectious Disease: Clinical Cases Uncovered. By H. McKenzie, R. Laing, A. Mackenzie, P. Molyneaux and A. Bal. Published 2009 by Blackwell Publishing. ISBN 978-1-4051-6891-5.

Table I Innate immunity and infection.

Function	Increased infection risk
Stomach acid	Patients on antacids have increased susceptibility to diarrhoeal pathogens
Respiratory tract mucociliary escalator	Patients with cystic fibrosis have recurrent chest infections
Skin	Serious burns often become infected

can live for years. Macrophages can be free in the tissues (e.g. alveolar macrophages) or fixed in the liver, spleen and lymph nodes as part of the **mononuclear phagocytic system**.

There are two separate components to the activity of phagocytic cells – firstly **phagocytosis** itself, which involves engulfing the microorganism and internalising it within the cell. The organism is then held within a 'phagosome'. The second process is **intracellular killing**, a complex process that involves fusion of the phagosome with lysosomes to form a 'phagolysosome', containing lysosomal enzymes which attack the organism. The energy-dependent production of free radicals, which are toxic to the organism, also takes place. Capsulate bacteria such as *Streptococcus pneumoniae* are resistant to phagocytosis unless they are 'opsonised' (see below). In contrast, organisms such as *Mycobacterium tuberculosis* are easily phagocytosed but are resistant to intracellular killing. The mononuclear phagocytic system is important in recognising and dealing with pathogens regionally throughout the body. Thus the spleen clears the blood, the liver the enterohepatic circulation and the regional lymph nodes drain peripheral sites.

Opsonisation greatly improves the efficiency of phagocytosis and requires the involvement of antibody and/or complement. Phagocytic cells have receptors for both antibody and complement-fixed components (see below), so if a microorganism is coated with antibody or complement it is much more easily phagocytosed. This 'coating' process is known as opsonisation and is particularly important for capsulate bacteria such as *S. pneumoniae* or *Haemophilus influenzae*.

Acquired immunity

Acquired immunity involves the creation of immunological memory so that a response to a particular antigen is 'learned' and will work faster and better the next time that antigen is encountered. Each microorganism contains a complex mixture of antigens and each antigen itself is (usually) a mixture of **epitopes**. It is at the level of the epitope that individual clones of immune cells recognise foreign material – each clone is unique and will only be stimulated to divide and expand if it encounters an epitope that matches its antigen recognition site. Acquired immunity has two major arms, **humoral immunity** (antibody and complement) and **cell-mediated immunity** (T cells and macrophages). It may be useful to revise some basic background and terminology with regard to white cells at this point (Table J).

Table J Basic haematological terms.

Polymorphs	• Neutrophils (important in acute inflammation) • Eosinophils (important in allergy and helminth infection) • Basophils
Lymphocytes	• B lymphocytes (site of maturation not clear) – precursors of plasma cells • T lymphocytes (mature in thymus) – further subdivided into CD4 (helper) and CD8 (suppressor/cytotoxic) cells
Leucocyte	Generic term for a white cell
Neutrophilia	Increase in circulating neutrophils
Lymphocytosis	Increase in circulating lymphoctyes

Humoral immunity

Antibody

Immunoglobulins (Ig) are the proteins that have antibody activity and five classes exist: IgM, IgG, IgA, IgE and IgD. If you were able to purify all of any one immunoglobulin class (e.g. IgG) from a single patient, it would contain antibody of a wide range of specificities against thousands if not millions of epitopes. IgM is the major part of the primary antibody response the first time a pathogen is encountered. Thus it is a valuable marker of current or recent infection in serological diagnosis, as outlined in Laboratory diagnosis of infection (Part 1). IgG forms the bulk of the secondary response once the IgM begins to diminish or on repeated exposure to the same antigen. Whatever immunoglobulin class it is, antibody is synthesised by plasma cells. A clone of plasma cells differentiates from a B lymphocyte after it has been stimulated by a specific **epitope**, although this process requires T cell help (see below).

Complement

The **complement cascade** is a complex series of reactions involving 20 or so proteins – the most important are numbered from C1 to C9. The combination of antibody (only IgG and IgM are active in this regard) and its specific antigen triggers a cascade of reactions, with complement being said to be 'fixed' by this process. Some bacterial products (e.g. endotoxin) are able to trigger the complement cascade directly in the absence of antibodies – not by the conventional route but by something called the 'alternative pathway'. This activity contributes to the pathophysiology of Gram-negative shock (see below). Whilst the immunological mechanisms involved are complex, it is possible to summarise the role of antibody and complement in infection in a relatively simple way (Table K).

Humoral immunity is important in extracellular infection as neither antibody nor complement can exert their effects inside the host cell. Thus this arm of the immune response is most commonly involved in response to bacterial infection. It is part of an acute inflammatory response and bacterial infection is therefore commonly accompanied by an increase in circulating neutrophils.

Table K Possible roles of antibody and complement in infection.

- Antibody neutralises bacterial toxins and prevents them from acting at their target site
- Antibody neutralises viruses in the extracellular viraemic stage, preventing them from infecting host cells
- Antibody prevents the adherence of microorganisms at epithelial surfaces
- Antibody opsonises capsulate organisms (*Streptococcus pneumoniae*, *Haemophilus influenzae*) making them easier for phagocytic cells to ingest
- Complement opsonises bacteria
- Complement lyses Gram-negative organisms by punching holes in the cell wall; this is particularly important in the defence against meningococci
- Many of the by-products of the complement cascade are chemotactic – they attract polymorphs to the site of the reaction where they can phagocytose the invading organism
- The antibody response provides the basis for serological diagnosis

Cell-mediated immunity

Cell-mediated immunity is a collaborative venture between T cells and macrophages. Macrophages 'present' antigen to T cells which then produce a range of small molecules called **cytokines** which control the immune response, both in terms of up-regulation (pro-inflammatory) and down-regulation (anti-inflammatory). There are various types of T cell:

- **CD4 helper cells**, which have two different functions:
  - Th1 cells support cell-mediated immunity. They activate macrophages to ingest and kill or contain the pathogen.
  - Th2 cells support B cells in producing an antibody response.
- **CD8 cells**, which are either suppressor or cytotoxic cells:
  - Suppressor cells help 'switch off' the immune response when it is no longer needed.
  - Cytotoxic cells kill infected host cells or cells detected as foreign.

The mechanisms by which the immune system 'decides' to respond to an antigen by going down the cell-mediated as opposed to the antibody production route, or vice versa, are not well understood. However, it is clear that cell-mediated immunity is important in defence against intracellular infection – this largely means viral infection, but includes certain bacteria such as *Mycobacterium tuberculosis* and fungi. Given the importance of lymphocytes in this process, it is not surprising that an increase in circulating lymphocytes (lymphocytosis) is characteristic of viral or intracellular infection.

Inflammatory markers of infection

In clinical practice, inflammatory markers provide evidence to help decide whether a patient has an infection and also whether the patient's condition is improving or deteriorating during the course of therapy. Assessing this kind of evidence longitudinally over the course of an illness or hospital stay is an important component of good clinical management. Other than regular white cell counts, C-reactive protein (CRP) is a commonly measured index of inflammation. CRP is produced by the liver as part of a cytokine-induced inflammatory response and is defined as an acute phase protein. Whilst CRP is not specific for infection, it is a marker of the inflammatory response and is often useful in monitoring the progress of antibiotic treatment. Inflammatory markers that might be considered in the diagnosis and management of infection are listed in Table L.

Table L Useful measurements in the diagnosis and management of infection.

Factors associated with infection	Comments
Elevated temperature	
Elevated CRP levels	Can be elevated in autoimmune conditions and post surgery
Elevated white cell count	Viral infection is (usually) accompanied by a lymphocytosis Bacterial infection is (usually) accompanied by a neutrophilia Helminth infection is (usually) accompanied by an eosinophilia

Shock

Whilst the immune system is generally thought of as protecting the host from infection, there are circumstances in which the immune response itself becomes part of the problem and has pathological consequences. Septic shock is an important example of this and this section provides an overview of a very complex topic.

Shock is defined clinically as acute circulatory failure with inadequate or inappropriate perfusion resulting in generalised cellular hypoxia. This may arise in several ways. A patient who loses litres of blood following trauma to a major blood vessel will develop **hypovolaemic shock** – hypoxia will result from inadequate perfusion of the tissues because there is no longer enough blood to go round. Alternatively, if there is some compromise to cardiac function (e.g. as a result of myocardial infarction) a patient might develop **cardiogenic shock**. In this scenario, there is still enough blood to go round, but there is inadequate perfusion because the heart does not pump it round efficiently enough.

Septic shock occurs when the immune response 'over reacts' to sepsis and a complex sequence of inflammatory events takes place which seriously affects the functioning of the cardiovascular system. This results in inadequate perfusion of the tissues and organs and the consequent hypoxia can cause organ failure.

Definitions

- **Systemic inflammatory response syndrome** (SIRS) is a widespread inflammatory response to a variety of clinical insults (not necessarily infection). It is recognised by the presence of two or more of the following:

Temperature	*>38 °C or <36 °C*
Heart rate	*>90 beats/min*
Respiratory rate	*>20 breaths/min or $PaCO_2$ < 4.3 kPa*
White blood cell count	*>12 × 10^9 cells/L or <4 × 10^9 cells/L*

- **Sepsis** is SIRS plus definitive evidence of infection, e.g. in bacteraemia, a positive blood culture.
- **Severe sepsis** is sepsis with evidence of organ hypoperfusion or dysfunction, e.g. renal function deteriorating.
- **Septic shock** is severe sepsis with hypotension (systolic blood pressure <90 mmHg) despite adequate fluid resuscitation.

Mechanisms of shock

The final common pathway of septic shock is a complex cascade of inflammatory mediators which impairs or disrupts the functioning of the cardiovascular system. This results in hypotension, tissue hypoxia and ultimately leads to organ failure. However, there is more than one way in which this can come about and the complexities of the pathophysiology are still not fully understood. There are two well recognised bacterial triggers: endotoxic and Gram-positive shock.

Endotoxic shock

Endotoxic shock is triggered by endotoxin (also called lipopolysaccharide (LPS)), which is a structural component of the Gram-negative cell wall. Endotoxin links with endotoxin-binding protein and the complex triggers CD14 receptors on macrophages and circulating monocytes. These release tumour necrosis factor α (TNF-α) and interleukin 1 (IL-1), which are pro-inflammatory cytokines that up-regulate the immune system to cause inflammation. Other pro-inflammatory cytokines involved in the process include IL-2, IL-6, IL-8 and interferon γ (IFN-γ). This 'cytokine shower' results in fever (TNF-α and IL-1 are potent pyrogens), increases in acute phase proteins (e.g. CRP) and increased vascular permeability and polymorph adherence. The latter makes the microvasculature 'leaky' and this compromises the perfusion of organs and tissues.

In addition, the coagulation cascade is inappropriately activated, producing microthrombi that block small blood vessels and lead to disseminated intravascular coagulation (DIC), with widespread bleeding as the clotting factors become used up. Complement is also activated, releasing chemotactic factors that attract neutrophils to the site and encourage a widespread inflammatory response, including mast cell degranulation. Nitric oxide synthase is induced, resulting in increased nitric oxide (NO) production. NO reduces vascular resistance (hypotension), blocks mitochondrial electron transfer (cell hypoxia) and depresses cardiac contractility. Thus hypotension and organ failure are the end results of this sequence of events.

Gram-positive shock

Gram-positive shock is less well understood and the mechanism by which some Gram-positive organisms produce shock has been discovered much more recently. Certain toxins of *Staphylococcus aureus* and *Streptococcus pyogenes* have been shown to have the unusual property of stimulating T cells directly by binding to T cell receptors in a manner that is independent of the antigen specificity of the T cell. Thus, while a specific antigen will stimulate a small number of T cell clones with closely related receptors, these particular toxins stimulate a wide range of T cells in a non-specific manner, leading to a massive release of pro-inflammatory cytokines. Gram-positive toxins with this property are called **superantigens**.

Clinical management

Management of the patient with septic shock is a complex task that requires urgent referral to an intensive care unit. It is therefore important that every doctor recognises the clinical features of shock and their importance. In simple terms, the approach is to treat the source of infection (which may or may not be obvious), and to support the failing organs/tissues pharmacologically or mechanically. Thus, in addition to antibiotic cover for the infection, common interventions might be targeted as shown in Table M.

Table M Clinical significance of septic shock for different organs.

Tissue/organ	Clinical presentation	Intervention
Lungs	Adult respiratory distress syndrome	Mechanical ventilation
Kidneys	Renal failure	Dialysis
Heart	Cardiogenic shock	Inotropic support
Blood	Disseminated intravascular coagulation	Fresh frozen plasma
Vasculature	Hypotension	Intravenous fluids

Immunisation

Immunisation provides the opportunity to artificially manipulate the immune system to help prevent infection or at least minimise its effects. There are two major ways in which it is possible to confer immunity: passive and active.

Passive immunisation

Passive immunisation involves the transfer to a susceptible host of pre-existing antibody taken from subjects who already have immunity to the organism in question. Passive immunity is conferred immediately but is relatively short lived and no immunological memory is created. In some instances, pooled human normal immunoglobulin is used as a mixture of antibodies from the 'average' population is likely to contain antibody to most common human pathogens. However, disease-specific immunoglobulin preparations are also available and are prepared from subjects with high titres of antibody to individual agents such as hepatitis B, varicella zoster virus, rabies and tetanus.

Active immunisation

Active immunisation involves stimulating the immune response by the administration of either the organism itself or an important antigenic part of it. In contrast to the passive form, active immunity takes some time to develop, but lasts longer and creates immunological memory. There are many different types of active vaccine, but the most important distinction is between live and inert vaccines. Live vaccines contain organisms whose pathogenicity has been attenuated in some way but still retain their antigenicity.

Live vaccines

Live vaccines multiply *in vivo* and therefore provide an ongoing stimulus to the immune response, unlike inert vaccines which provide a 'one off' stimulus. Thus live vaccines may produce a better response, or may not have to be given so often. However, one disadvantage is that even an attenuated vaccine can cause disease in a patient with a diminished immune response. It follows that live vaccines are contraindicated in pregnant women and in inmunosuppressed patients, e.g. acquired immune deficiency syndrome (AIDS) patients. Live vaccines are most commonly used for viral infections, e.g. MMR (measles, mumps and rubella), but one of the oldest bacterial vaccines – BCG (Bacille Calmette-Guérin) – is an attenuated form of *Mycobacterium bovis*.

Inert vaccines

Inert vaccines may take various forms and research is generally directed towards finding a single antigen that results in a protective immune response against the whole organism. The simpler and more defined the vaccine, the less the chance of serious side effects, e.g. toxin-mediated disease is most effectively countered by immunisation against the specific toxin itself. The antigen is a **toxoid** (i.e. toxin-like) – the antigenicity of the toxin is retained but the molecule is altered in such a way that it no longer has any toxigenic activity. Once a specific antigen has been identified, it can be produced by extraction from cultures of the organism, or sometimes more easily by genetic engineering. Thus hepatitis B surface antigen is produced not from viral cultures, but by cloning the gene concerned into yeasts and 'tricking' the yeast into producing large quantities of hepatitis B surface antigen.

Conjugate vaccines

It has long been recognised that the immune response to carbohydrate antigens is poor, especially in the very young. This is because carbohydrates often consist of identical repeating sugar residues and the immune system finds it more difficult to deal with a single repeating epitope than with the variety of epitopes found in protein molecules. The introduction of conjugate vaccines, in which carbohydrate antigens are covalently linked to a protein, has been a major advance in protection against infection by bacteria with carbohydrate capsules, e.g. *Haemophilus influenzae* type b (Hib vaccine), *Neisseria meningitidis* group C (Men C) and *Streptococcus pneumoniae* (Prevenar).

Approach to the patient

As with any branch of medicine, the key to making a diagnosis of infection in most patients is a well taken history. It is not the remit of this book to teach students about the fundamentals of history taking but there are certain aspects that are of particular significance when approaching a patient with suspected infection. Remember that no two patients are identical and a different approach is often needed for different situations. No infectious diseases book would be complete without highlighting the need for adequate **handwashing** before and after you approach the patient.

History

Introduction

After introducing yourself and explaining your intentions, the following details should be confirmed: patient's age, date of birth, current and previous occupations, where they currently live and where they come from originally. While this gives useful information, it also allows you to build up a rapport with the patient before enquiring about their main complaint.

Does my patient require isolation?

There are two kinds of isolation – **protective isolation**, when a neutropaenic patient is protected from the everyday microorganisms that inhabit the outside world, and **source isolation**, when it is necessary to prevent the spread of infection from the patient to the outside world, or at least other patients and staff. These forms of isolation are summarised in Table N. Before starting to take the history from the patient, you need to ask yourself, 'Does the patient have an infection that could be transmitted to you, to other staff, or patients?' A clue to this might be given in the doctor's referral letter or the risk may become obvious to you at some stage during the consultation. Some examples of infections that require isolation are immediately evident – chickenpox is an obvious one. Others may be a bit more subtle, e.g. the patient with an unresolving pneumonia who might be harbouring tuberculosis. Some examples of conditions that require isolation are given in Table O.

Whilst most patients do not require any special precautions, some (e.g. methicillin-resistant *Staphylococcus aureus* (MRSA) colonised patients) will require barrier nursing, ideally in a single room. For some respiratory pathogens (e.g. multidrug-resistant tuberculosis), negative pressure isolation facilities are appropriate and staff may need to wear face masks until the infectivity of the patient has been established. If a patient requires isolation, it can be a very frightening experience for them so you need to spend time sensitively explaining what is involved and why.

Presenting complaint and history of presenting complaint

Remember to record the presenting complaint in the patient's own language, e.g. 'I have been feeling hot and sweaty for the last three days' or 'I feel as though I have had the flu'. Many symptoms due to infection are non-specific – headache, fever, rigors, chills, sweating, malaise, fatigue, aching limbs, backache, vomiting and diarrhoea. Make sure that when you have elicited the presenting complaint, you go on to find out what your patient understands by the terms they have used. For some patients, diarrhoea is one bowel movement a day, whilst for others this might be regarded as constipation. A list of terms that may require clarification is given in Table P. Sometimes the presenting symptoms are relatively 'organ-specific' such as purulent sputum or dysuria, but in other cases you will need to spend time running through a systematic inquiry to be sure that nothing important has been left unreported by your patient.

Infectious Disease: Clinical Cases Uncovered. By H. McKenzie, R. Laing, A. Mackenzie, P. Molyneaux and A. Bal. Published 2009 by Blackwell Publishing. ISBN 978-1-4051-6891-5.

Table N Forms of isolation.

Type of isolation	Features
Protective isolation (reverse barrier nursing)	• Physical separation of a patient at high risk (e.g. neutropaenic) from common organisms carried by others • Single room with door closed ± positive pressure ventilation • Contact between staff and patients kept to a minimum to reduce risk of infection passing to patient • Staff to wear protective clothing – disposable aprons and gloves – and use strict hand hygiene. Keep charts outside room to reduce risk of introducing infection to patient
Source isolation (barrier nursing)	• Physical separation of one patient from others, in order to prevent spread of infection • Single room with door closed ± negative pressure ventilation • Contact between staff and patients kept to a minimum to reduce risk of infection to staff or via staff to other patients • Staff to wear protective clothing – disposable aprons and gloves – and use strict hand hygiene. Keep charts outside room to reduce risk of contamination from patient

Table O Examples of infections that require source isolation of the patient.

Infection	Comments
Antibiotic-resistant organisms	Patients with MRSA, extended spectrum β-lactamase (ESBL) or vancomycin-resistant enterococci (VRE) infections (see Table G, Antimicrobial chemotherapy, Part 1) would normally be placed in source isolation to prevent spread
Chickenpox or shingles (varicella zoster virus, VZV)	Infectious until all lesions are dry and crusted
Clostridium difficile associated diarrhoea (CDAD)	Common healthcare-acquired infection. Spores from organisms contaminate the environment. Spores on hands are not killed by alcohol rubs – handwashing is required
Gastroenteritis	All patients should be treated in single rooms and barrier nursed, or cohorted in the same ward or ward bay if single rooms are not available. Note that some organisms (e.g. norovirus) are more infectious than others (e.g. *Campylobacter*)
Impetigo	Staphylococcal skin infection that spreads easily to other patients
Measles	Healthcare workers should not enter the patient's room unless known to be immune
Meningococcal disease	Prophylaxis is given to close contacts. Healthcare workers do not usually come into this category unless for example they have given mouth to mouth resuscitation
Bordetella pertussis (whooping cough)	In outbreaks, patients are sometimes cohorted in the same room
Pulmonary tuberculosis (TB)	If there is a risk of multidrug-resistant TB, patients should be isolated in negative pressure rooms. Otherwise smear positive patients (i.e. acid and alcohol fast bacilli (AAFB) seen in sputum) should be isolated until they improve on treatment and have negative microscopy
Viral haemorrhagic fever	Rare infections (e.g. Lassa fever) which require immediate involvement of the health protection authorities, strict isolation and total body protection for healthcare workers

Table P Questions relevant to common presenting complaints.

Presenting complaint	Ask about
Diarrhoea	Duration, frequency, accompanying symptoms (abdominal pain, fever, vomiting), foods consumed prior to onset, recent antibiotics, family history of inflammatory bowel disease or coeliac disease, recent travel
Headache	Neck stiffness, sensitivity to light, vomiting, rash, speed of onset of headache (very sudden onset suggests intracranial bleed rather than meningitis)
Fever	Duration, pattern (does it occur every day?), accompanying symptoms, any rigors or shivers, drug history (some drugs cause fevers), recent travel
Rash	Duration of rash and where it first appeared, distribution, accompanying symptoms, blanching or non-blanching, does it itch, drug history, contact history (e.g. chickenpox or impetigo). Has the rash evolved?
Sore throat	Is the pain worse on one side (may indicate a peritonsillar abscess), difficulty swallowing, difficult or noisy breathing (listen for stridor as an indicator of airways obstruction), lymph node enlargement (common with glandular fever)

Table Q An example history.

- Your patient is Mr Peter Bell, a 27-year-old accountant
- *Presenting complaint*: rash and fevers
- *History of presenting complaint*:

When did the rash begin?	Three days ago
Where did it start?	I first noticed it on my chest but by the next day it had spread to my face, then my arms and legs
Any other symptoms?	I have been feeling hot and cold since the rash began, in fact I thought I was coming down with the flu at first but then I noticed the spots
Is the rash itchy?	Very itchy, it is driving me mad!
Are you still getting more spots?	I have had new spots every day since it started
Have the spots changed?	Yes, the spots were initially like little blisters with clear fluid then they became filled with pus and looked different from the new spots. After a couple of days they scab over

- *Remainder of history*: this reveals that your patient has previously been in good health and takes no medication. He has never used any illicit drugs and was last abroad 3 years ago when he visited France. He lives with his wife and 2-year-old son. His son had a similar rash 2 weeks ago but it was much milder that Mr Bell's
- *On examination*: the rash is made up of discrete spots about 0.5 cm in diameter. Some of them are fluid filled (vesicles) or have pus in them (pustules). The original spots have scabbed over. The distribution is wide with most of them over the chest, abdomen and head, but no part of the body is spared. You also notice some spots in the oral cavity including the soft palate

The history of the presenting complaint may be obvious, e.g. a patient with cellulitis may report having had a painful red leg for 4 days. In some cases, the history of the presenting complaint is more involved and requires you to ask specific relevant questions. An example history of a patient with chickenpox is given in Table Q.

Other aspects of the history of particular relevance to infectious diseases

Having obtained details of the presenting complaint, the remainder of the history should be obtained in the usual fashion. However, there are some parts of the history that are of particular importance in infectious diseases.

Table R Examples of some infections that may be associated with travel.

Disease	Incubation period	Time since travel to presentation	Usual symptoms
Entamoeba histolytica	2–4 weeks	1–12 months	Bloody diarrhoea (presents early) or liver abscess (may present as abdominal pain and fever after several months)
Falciparum malaria	8–25 days	Up to 6 weeks	Fever, flu-like symptoms
Vivax malaria	8–27 days	Up to 1 year	Fever, flu-like symptoms
Dengue fever	3–15 days	Up to 3 weeks	Fever, headache, rash
Giardiasis	7–28 days	Up to 1 month	Diarrhoea
Hepatitis A	15–50 days	Up to 6 weeks	Jaundice
Cutaneous leishmaniasis	4–6 weeks	1–3 months	Non-healing skin lesions
Pulmonary tuberculosis	6–24 days	Weeks to months	Cough, fevers
Typhoid fever	10–14 days	Up to 3 weeks	Fever, headache, dry cough

Travel history

A travel history is essential in all patients. Bear in mind that as some travel infections may become apparent months after returning from an endemic region, you should ask about travel in the last 12 months at least (Table R). As well as places visited, you need to find out about the type of accommodation, activities while there and any immunisation/prophylaxis history.

Social history

Many medical students and doctors take a cursory social history, largely focusing on smoking and alcohol consumption. This approach runs the risk of missing out on some very important information. For example, a sexual history is obviously needed if a sexually transmitted infection is being considered and a history of injection drug use is relevant to many infections including blood-borne viruses, endocarditis and skin abscesses. The social history should include hobbies and recreations, pets/other animal contacts and an immunisation history. Be sure to ask about contact with anyone who has had similar symptoms.

The occupational history is important, both as a risk factor for developing an infection (e.g. forestry workers and Lyme disease) and for infecting others (e.g. a teacher with tuberculosis).

Drugs and allergies

A list of current medication should always be recorded as part of the history. If patients cannot recall their medicines, you should contact their GP surgery for details. Take particular note of immunosuppressive therapy that can increase the risk of infection. Drugs that reduce stomach acid, such as proton pump inhibitors or ranitidine, can increase the risk for gastroenteritis. Diuretics or vasodilators (often given to patients with cardiac disease) can worsen hypotension/shock in the septic or dehydrated patient. Remember that any patient with diarrhoea should be asked about antibiotic treatment in the previous month – if an antibiotic has been prescribed that makes the possibility of *Clostridium difficile* diarrhoea more likely.

Antibiotics are amongst the commonest type of drug to which patients will report an allergy. Be sure to find out what they mean when they say that they are allergic – some patients simply mean that the drug makes them a bit nauseated, rather than causing a true allergic reaction. About 10% of people claim to be penicillin allergic, but only 1% truly are. A true allergy may be an immediate anaphylactic reaction or a delayed hypersensitivity reaction (rash after 2–3 days). About 5–10% of those allergic to penicillin are also allergic to cephalosporins.

Infection emergencies

Ideally, the history should be obtained from the patient in a systematic fashion and in a calm environment. However, some patients may present *in extremis*, e.g. unconscious or drowsy, having seizures or hypotensive. Conditions such as meningitis, septicaemia and encephalitis may present in this way. The priority in these situations is patient resuscitation and stabilisation, but once this is done you should not forget that valuable information can often be obtained from a friend or relative (over the phone if necessary).

Examination

The end of bed assessment gives you an indication of whether the patient is alert, orientated and generally well, or drowsy, confused and obtunded. Record the patient's pulse, blood pressure, respiratory rate, oxygen saturation and conscious level at the beginning of the examination. The patient's systemic inflammatory response syndrome (SIRS) can be given a score of 0–4 from the pulse, temperature, respiratory rate and white cell count (Table S). A SIRS score of 2 or more is in keeping with an inflammatory response – this is due to infection in many (but by no means all) cases.

General examination

As with all patients, a full examination is appropriate in the patient with infection. Although the examination is sometimes focused on an obvious site of infection (e.g. leg cellulitis), many patients have a non-specific history so that a full examination is especially important. Do not restrict your examination to the major organs, remember that clues about the diagnosis may be gleaned from a wider examination including the hands (finger clubbing), oral cavity (thrush or oral ulceration), sclera (jaundice), genitalia (ulceration or discharge) and skin, including scalp (petechial rash).

Table S Scoring system for SIRS. Two or more points are indicative of an inflammatory response.

Feature	Abnormality	Score (points)
Temperature (°C)	>38 or <36	1
Pulse (beats/min)	>90	1
Respiratory rate (breaths/min)	>20	1
White blood cell count ($\times 10^9$/L)	>12 or <4	1

Making a diagnosis

The history alone will allow you to make a diagnosis in up to 90% of cases. The physical examination should help to confirm the diagnosis or perhaps make one of the differential diagnoses more likely than the others.

In considering a differential diagnosis, remember that 'common things are common'. The epidemiology and clinical features of most common infections are quite well known. Therefore the patient is not likely to have a tropical condition if they have not travelled abroad.

It is likely that you will want to perform a number of investigations and this aspect of the patient's assessment is covered in the cases that follow. In general, *specific tests* aimed at confirming a suspected diagnosis tend to be of more value than a broad 'infection screen'.

If the patient is to be tested for a serious communicable disease, e.g. hepatitis B, hepatitis C, human immunodeficiency virus (HIV) or tuberculosis, there should be pre-test discussion regarding the condition, the nature of the test and the implications of a positive or negative result.

Finishing the consultation

Before finishing the consultation, invite the patient to take a moment to think of anything else that might be of relevance. It is sometimes the case that the process of giving their history will prompt them to think of other symptoms or events that they had not told you about earlier.

Once you are satisfied that you have all the relevant information, you should explain to the patient the likely diagnosis, planned investigations and treatment. Sometimes it may not be possible to give a diagnosis at this stage and the reasons for this should be explained. For patients who are critically ill, you will need to speak to their next of kin about their condition and prognosis. Where possible, this should be with the patient's consent.

Finally, ask yourself the question, 'Who do I need to tell about this patient?' This could include senior staff on your own team or specialists in other areas, e.g. intensive care if ventilation might be required or renal physicians if dialysis might be required. If there is a suspicion that the patient may be infectious to others, you may need to advise the hospital infection control team, and if there are wider community implications you may have to inform the local health protection team (public health).

Table T Examples of notifiable diseases.

Notifiable disease	Causative organism
Anthrax	*Bacillus anthracis*
Cholera	*Vibrio cholerae*
Legionellosis	*Legionella pneumophila*
Measles	Measles virus
Tuberculosis	*Mycobacterium tuberculosis*
Typhoid	*Salmonella typhi*

In the short term, you may need to advise any potential contacts about the risks and explain to the patient and family the need to apply infection control precautions and possibly even restrict visiting.

Notifiable diseases

Some infections are required to be notified to the Health Protection Agency or equivalent local health protection department (Table T). This is a mechanism that can help to identify outbreaks or unexpected trends in infectious diseases. Microbiology departments will inform the relevant agency if they have identified a notifiable pathogen in the laboratory, but remember that some diagnoses are made clinically and laboratory confirmation is not always possible (e.g. meningococcal infection diagnosed on the basis of the history and a typical rash, but with negative cultures due to antibiotics administered before admission). In these circumstances the doctor in charge of the patient should inform the local department or Health Protection Agency directly. It does no harm if the health protection team hears about a patient from more than one source!

Case 1 A 68-year-old woman with bloody diarrhoea

A 68-year-old woman was referred to the accident and emergency department with a history of abdominal pains and diarrhoea. She had noticed blood in the stools.

What differential diagnosis would you think of at this stage?

1 Gastroenteritis.
2 Diverticulitis.
3 Inflammatory bowel disease.
4 Colonic carcinoma.
5 Ischaemic colitis.
6 Pseudomembranous colitis.
7 Bleeding haemorrhoids.

What details would you like to elicit from the history?

Presenting complaint and history of presenting complaint

You would want to know more about the diarrhoea:

- What is its frequency – day and night?
- Is the blood mixed with or separate from stool?
- Has she had similar symptoms previously?
- Was the onset abrupt or gradual?
- What is its duration?
- Are any other family members or contacts symptomatic?

Associated symptoms

Does the patient have any of the following?

- Abdominal pain.
- Fever.
- Weight loss.
- Tenesmus.
- Vomiting.
- Any eye, skin or joint symptoms (associated with inflammatory bowel disease).

Past medical history/drug history

- Is there a history of vascular disease?
- Is there a history of treatment with any of the following:
  - Anticoagulants.
  - Antiplatelet agents.
  - Antimicrobials – recent or current.

Family history/social history

- Is there a family history of inflammatory bowel disease?
- Occupation.
- Travel history.

Further history reveals that the diarrhoea was of abrupt onset with associated severe abdominal pain and tenesmus. The patient was passing small bloody stools once or twice an hour at the time of admission. She described fevers and anorexia but no vomiting. She is a farmer's wife, living on a cattle farm. She had a past history of hypertension for which she took atenolol. She has not been prescribed any other medication in the last 6 months. She had no family history of bowel disease. She had travelled to Egypt on a family holiday from which she had returned 2 weeks before taking ill.

Review your differential diagnosis

1 *Gastroenteritis.* This would seem likely given the abrupt onset and short history. Blood in the stools is more likely with some infections, e.g. *Shigella* and *Escherichia coli* O157:H7, than with others, e.g. viral gastroenteritis or *Salmonella*. *Shigella* infections are uncommon in the UK, but the patient has been to Egypt recently. Since the onset was 2 weeks after return, this makes shigellosis unlikely.

Infectious Disease: Clinical Cases Uncovered. By H. McKenzie, R. Laing, A. Mackenzie, P. Molyneaux and A. Bal. Published 2009 by Blackwell Publishing. ISBN 978-1-4051-6891-5.

Table 1.1 Common causes of infective diarrhoea in the UK.

Bacteria	Viruses	Protozoa
Campylobacter spp.	Rotavirus	*Cryptosporidium parvum*
Salmonella spp.	Norovirus	*Giardia lamblia*
Shigella spp.	Adenovirus	
E. coli O157		
Clostridium difficile		

The commonest bacterial causes of diarrhoea in the UK are *Campylobacter* and *Salmonella. E. coli* O157 is less common, but it is clinically important because of its complications and does present with bloody diarrhoea (Table 1.1). A further possible cause of bloody diarrhoea in a patient with a recent history of travel to the tropics is amoebic dysentery. Special arrangements with the laboratory for examination of a 'hot stool' by microscopy to look for amoebic trophozoites would be required for diagnosis.

2 *Diverticulitis.* This can be associated with bleeding per rectum but the frequency of the diarrhoea would make diverticulitis less likely.

3 *Inflammatory bowel disease.* This would be an equally plausible diagnosis to gastroenteritis. If the history was of more than 2 weeks' duration, inflammatory bowel disease would become the favoured diagnosis. The lack of a family history or previous episodes does not eliminate this diagnosis.

4 *Colonic carcinoma.* Rectal bleeding should always raise the possibility of colonic carcinoma but the frequency of the bloody stools and the associated symptoms would make this less likely than options 1 or 3 above.

5 *Ischaemic colitis.* This can often be hard to diagnose. It is more likely in older patients with a history of vascular disease or atrial fibrillation but needs to be considered in patients with bloody stools and abdominal pain.

6 *Pseudomembranous colitis* or *Clostridium difficile-associated diarrhoea (CDAD).* It is not possible to differentiate pseudomembranous colitis from other causes of colitis on clinical criteria alone. However, the lack of recent antibiotic treatment would go very much against pseudomembranous colitis in this case.

7 *Bleeding haemorrhoids.* The other features in this case would make local bleeding from haemorrhoids most unlikely.

What next?

Examination – what are the key things to look for?

General examination

Is there a fever or other features of an inflammatory response (see SIRS in Infection and immunity, Part 1)? This could suggest infection or inflammation. Remember in this case, atenolol may have prevented development of a tachycardia as part of her inflammatory response.

Hydration

The clinical features of dehydration include loss of skin turgor, dry mucous membranes, poor peripheral perfusion, tachycardia and a low blood pressure. A significant postural fall in blood pressure (BP) – more than 20 mmHg fall in systolic pressure when going from lying to standing – is an earlier sign than reduced blood pressure.

Abdominal examination

- Look for abdominal distension – suggesting bowel obstruction or dilatation.
- Feel for abdominal tenderness – is it localised or diffuse? Localised tenderness might favour a particular diagnosis, e.g. left iliac fossa tenderness is typical in ischaemic colitis or diverticulitis.
- Listen for bowel sounds – these may be increased in obstruction or gastroenteritis. Absent bowel sounds can develop in peritonitis, which might develop as a consequence of ischaemic colitis.
- Examine the rectum for blood or masses.

On examination the patient appeared in discomfort. She had a temperature of 37.5 °C, pulse 65 beats/min and BP 115/70 mmHg. There was no fall in BP on standing. The abdomen was diffusely tender but not distended. There were bowel sounds present. The rectum was empty with blood staining of the examining finger.

What would you do now?

Fluids

Given the nature of the patient's problem and the wide differential, the first thing to do is to provide rehydration. This would best be done using intravenous (IV) fluids – 0.9% saline would be most appropriate. By keeping the patient hydrated using IV fluids, you can allow her to remain fasted in case you subsequently diagnose a surgical problem that will require her to go to theatre in the near future.

Bloods

Take blood for urea, electrolytes, creatinine and a full blood count. Send blood cultures to microbiology in light of the fever and a sample to the blood transfusion laboratory for 'group and save' in case the blood loss proves sufficient to require transfusion.

Imaging

A plain film of the abdomen would be useful to look for bowel dilatation. Dilatation of the large bowel can occur in severe colitis, e.g. due to inflammatory bowel disease – so-called 'toxic dilatation'.

Cultures

Stool cultures are pivotal in differentiating between infective and non-infective causes of bloody diarrhoea. Specimens are routinely cultured for *Campylobacter* spp., *Salmonella* spp., *Shigella* spp. and *E. coli* O157. *Campylobacter* spp. are by far the commonest bacterial cause of diarrhoea, followed by *Salmonella* spp. Remember that in some clinical scenarios, you would request tests other than culture on a stool sample, e.g. enzyme immunoassay for *Clostridium difficile* toxin in suspected pseudomembranous colitis, and microscopy for parasites, cysts and ova (P,C&O) in suspected parasitic infection. For amoebic dysentery, a 'hot stool' would be examined microscopically for the presence of *Entamoeba histolytica* vegetative forms ingesting red cells. Cysts may also be seen, more commonly in an asymptomatic patient. Remember that liver abscess is an important long-term complication of amoebic dysentery. A viral cause is less likely in bloody diarrhoea, but remember that you may have to separately request tests for viral causes of diarrhoea.

Some results are back – how do you interpret them?

The initial blood tests have returned showing the following results:

		(Normal range)
Haemoglobin	*11.7 g/dL*	*(11.5–16.5)*
White blood cell count	21.5×10^9*/L*	*(4–11)*
Platelet count	170×10^9*/L*	*(150–400)*
Serum sodium	*133 mmol/L*	*(137–144)*
Serum potassium	*3.2 mmol/L*	*(3.5–4.9)*
Serum urea	*8.2 mmol/L*	*(2.5–7.5)*
Serum albumin	*38 g/L*	*(37–49)*

The high white cell count would support an infective or inflammatory cause for the bloody diarrhoea. Despite the blood loss, her haemoglobin is still normal so there is nothing in the full blood count to suggest a large volume of bleeding from the gut. The urea is slightly raised in keeping with diarrhoea-induced dehydration and the low potassium is also likely to be due to losses incurred through diarrhoea. The albumin level is normal – a chronic inflammatory process would be likely to cause a reduced serum albumin.

What happened next?

The patient was treated with IV saline with potassium. Over the next 24 hours she passed 15 small volume, bloody stools. On day two of admission the laboratory phoned to inform the ward that E. coli O157:H7 had been isolated from a stool sample. This organism is shed by cattle, so the patient's place of residence may well have been of relevance to her illness.

Should this change your treatment?

The treatment should stay the same since antibiotics are not used in *E. coli* O157 infection. In fact, studies suggest antibiotics may actually cause some harm by increasing the risk of complications. She needs to be kept in isolation in the hospital as *E. coli* O157 is highly infectious and might be transmitted to other patients. The health protection department ought to be informed of the diagnosis so that they can look for the source of infection and try to prevent other cases.

This patient also needs to be monitored for the development of the complication associated with this infection – haemolytic uraemic syndrome (HUS). The components of this are obvious from the name – haemolysis of red blood cells (Plate 8) and uraemia due to renal failure. *E. coli* O157 is the commonest cause of renal failure in children in the UK. Older patients are also at increased risk of HUS.

Escherichia coli O157:H7 and HUS

HUS arises as a result of the effect of cytotoxins that are released from *E. coli* O157:H7. The toxins are known as verotoxins or 'shiga-like' toxins because of their similarity to toxins produced by *Shigella dysenteriae*, and this organism is probably the source of the toxin genes now found in *E. coli* O157. The same toxins can be produced by other O groups of *E. coli* but these may need specialist laboratory techniques to detect them. Such strains are generically referred to as enterohaemorrhagic or verocytotoxic *E. coli* (EHEC or VTEC). The O antigen on which *E. coli* classification is traditionally based is determined by the antigenic properties of the polysaccharide component of lipopolysaccharide (LPS), which is an intrinsic component of the cell wall.

Note that *E. coli* O157 strains are not found in blood – *E. coli* in a blood culture generally has nothing to do with gastroenteritis and usually arises from infection in the urinary tract or intra-abdominal sepsis, e.g. in the biliary tract. O157 strains are clinically different from the 'ordinary' strains of *E. coli* that form part of the normal flora in the gut. It is the cytotoxin, not the organism, that gets into the blood stream from the gut and causes renal problems and intravascular platelet aggregation. The platelet clumps cause red cell damage – hence haemolysis.

What are the features of HUS?

Many patients have few or no new symptoms when HUS first develops, which is why it is so useful to monitor blood tests. You would specifically want to detect:

- Rising blood urea.
- Falling platelet count.
- Evidence of haemolysis including:
  - Falling haemoglobin.
  - Elevated serum lactate dehydrogenase.
  - Raised bilirubin.
  - Fragmentation of red blood cells seen on the blood film.

As HUS worsens, patients may become aware of diminishing urine volumes, fatigue and mild jaundice.

What can be done to treat HUS?

If a patient develops HUS, supportive treatment with renal dialysis may be needed until the renal function recovers. Blood transfusion may also be required and some patients are treated with plasma exchange to try and remove the cytotoxin that is causing the problem.

The patient in this case remained free from complications and her diarrhoea settled after 4 days in hospital. Her case was referred to the local health protection department who arranged to inspect the farm for evidence of contamination of her water supply by E. coli O157:H7 and screened family contacts.

CASE REVIEW

A 68-year-old farmer's wife presented with sudden onset of severe abdominal pain and bloody diarrhoea. The differential diagnoses of bloody diarrhoea, including non-infective causes such as inflammatory bowel disease, were reviewed. Stool cultures revealed *E. coli* O157 as the cause, so blood counts and renal function were monitored, but fortunately she did not develop HUS.

KEY POINTS

- Diarrhoea is not always due to infection. It can be a non-specific symptom of sepsis in systems other than the gastrointestinal tract.
- Most gastroenteritis can be treated with rehydration alone, whether oral or intravenous. Antibiotics are not normally needed.
- In certain circumstances, however, antibiotics should be considered in addition to rehydration, e.g. HIV disease, neutropenia, severe sepsis, severe dehydration, prosthetic heart valves, malignancy.
- *Campylobacter* and salmonellae are the commonest bacterial causes of gastroenteritis in the community, with *Clostridium difficile* an important cause in hospital. Norovirus and rotavirus are common viral causes.
- *E. coli* O157:H7 is relatively uncommon but important due to the potential sequelae of HUS. It should not be treated with antibiotics as this may increase the risk of HUS.

Further reading and information

Health Protection Agency up-to-date information on food poisoning, epidemiology, guidelines and advice for healthcare workers and the public is available at http://www.hpa.org.uk.

Case 2 A 73-year-old man who has been feeling generally unwell for 2 weeks

A 73-year-old man is admitted to the hospital with a history of fever for the past 2 weeks. He also complains of weakness, anorexia and being 'generally unwell'.

Think about the common things first!

What are the common causes of fever in the elderly?

- Urinary tract infection.
- Lower respiratory tract infection.

How do you find the source of infection?

- Take a detailed history.
- Carry out a general and systemic examination.
- Perform 'relevant' investigations.

History

The patient is a smoker and is known to be suffering from chronic obstructive pulmonary disease. He has noticed an increase in severity of cough over the last few weeks. He appears alert. He does not give a history of increased frequency or burning on micturition.

Examination

- *General examination.* His temperature is 37.9 °C, pulse 98 beats/min and regular, BP 104/86 mmHg and respiratory rate 17 breaths/min. There is finger clubbing. There is no cyanosis, pallor or icterus. He does not appear unwell.
- *Systemic examination.* A few crackles are heard at the bases of both lungs. There is a pansystolic murmur heard at the apex. Abdominal examination reveals normal bowel sounds, no tenderness and no organomegaly.

Infectious Disease: Clinical Cases Uncovered. By H. McKenzie, R. Laing, A. Mackenzie, P. Molyneaux and A. Bal. Published 2009 by Blackwell Publishing. ISBN 978-1-4051-6891-5.

What baseline investigations would you like to do?

- Full blood count and biochemistry (urea and electrolyte, liver function tests).
- Blood cultures.
- Any other relevant cultures – perhaps sputum and urine culture?
- A chest X-ray.

Would you like to start him on medication?

Do we know what we are treating? Is the patient acutely unwell? The answer to both these questions is 'No', so there is no rush to start antibiotics. A diagnosis of bacterial endocarditis should be considered in any patient with fever and a heart murmur and it is particularly useful in endocarditis to make a bacteriological diagnosis before commencing antibiotic treatment. Obviously the need for immediate treatment depends on your clinical assessment of how acutely ill the patient is. Undue haste in starting antibiotic treatment may be counterproductive as it may render subsequent blood cultures sterile, putting you in the position of treating on empirical grounds without positive cultures. However, you must ensure that your patient is monitored in case his condition deteriorates.

Results of investigations

The laboratory has issued a microscopy result for sputum while awaiting results of culture. There are scanty polymorphonuclear cells and moderate numbers of Gram-positive cocci and short Gram-negative bacilli. There are no pus cells in the urine and no evidence of microscopic haematuria.

		(Normal range)
Haemoglobin	*12.1 g/dL*	*(13–18)*
White blood cell count	*13×10^9/L*	*(4–11)*

Serum C-reactive protein (CRP)	*223 mg/L*	*(<10)*
Liver function tests	*Normal*	
Serum creatinine	*127 μmol/L*	*(60–110)*
Serum urea	*5.1 mmol/L*	*(2.5–7.5)*

What do you do now?

The preliminary results suggest the possibility of an ongoing infection. Note the elevated white cell count and CRP. There are only scanty polymorphonuclear cells in the sputum and the urine appears normal. It may be a good idea to await further results while continuing to monitor the patient.

Later that afternoon, the microbiologist phones to say that both the blood culture bottles are positive and show Gram-positive cocci on microscopy. The microbiologist thinks that they are likely to be streptococci (see Fig. A, Laboratory diagnosis of infection (Part 1) for a revision of basic Gram stain reactions), but it will be a further 24 hours before the organism has grown on culture and can be fully identified.

How do you interpret this result – is it significant and if so what should you do now?

There are two questions to consider when you get a positive blood culture result:

1 Is the isolate a contaminant from skin?
2 If it is not a contaminant, what is the source of the bacteraemia?

Is it a contaminant?

Contaminants are picked up from skin, and this is more likely if there was inadequate skin preparation before taking the culture. Coagulase-negative staphylococci are the commonest contaminants, and are seldom significant in the absence of any long lines (e.g. a central venous pressure line) or prosthetic devices (e.g. heart valves). Note that coagulase-negative staphylococci are good at attaching themselves to such foreign bodies and causing infection. In addition, contaminants are commonly found in only one bottle of a blood culture pair. Since both bottles are positive in this patient, the culture is more likely to be significant. However, you cannot tell the species of the organism (i.e. what kind of *Streptococcus* it is) from the Gram stain, so we do not yet know for certain that both bottles are growing the same organism. Now that you have decided this is unlikely to represent contamination, you need to consider the second question.

What is the likely source of the bacteraemia?

If you treat a bacteraemia with short-term antibiotics it may well resolve temporarily, but then recur if you have not identified and treated the source of the infection. You may get clues to the likely source from the history, your findings on examination and from the type of organism isolated. The need for further investigations may become clear at this point.

Streptococci in blood cultures are not commonly contaminants though they can often cause a transient bacteraemia. The genus *Streptococcus* contains several species of clinical importance and some are associated with very distinctive clinical features. They are initially classified by their ability to haemolyse blood agar (see Fig. A Laboratory diagnosis of infection, Part 1 and Plate 5). There are quite a few different β-haemolytic streptococci of clinical significance, but *S. pyogenes* (group A strep) is particularly important. It can cause wound infections and local infections such as tonsillitis and pharyngitis, but it is well recognised as a cause of serious systemic infections such as necrotising fasciitis and fulminant shock with accompanying bacteraemia secondary to infection in other sites. Alpha-haemolytic streptococci are subdivided into *S. pneumoniae*, also called the pneumococcus and a common cause of pneumonia and meningitis, and a whole range of different species collectively referred to as viridans streptococci, which are the commonest cause of bacterial endocarditis.

What is the diagnosis?

The patient has a systolic murmur and an ongoing cough that has increased in intensity. Could this be pneumococcal pneumonia in a patient with a history of chronic obstructive pulmonary disease? Could this be endocarditis caused by one of the streptococci belonging to the viridans group?

The chest X-ray shows no evidence of consolidation. On the morning ward rounds, your consultant asks you to further investigate the possibility of a diagnosis of endocarditis.

Which investigations do you need to perform?

Establishing the diagnosis

The heart and its appendages are easily viewed by an echocardiogram. A transthoracic echocardiogram (TTE) is a non-invasive procedure that defines both the structure and the function of the heart. Unfortunately, TTE is not as sensitive as a transoesophageal echocardiogram (TOE). The heart is more clearly visualised by placing the transducer in the oesophagus. TOE is therefore invasive, and patients need to be sedated prior to the procedure.

The echocardiogram shows a small mobile mass on the posterior leaflet of the mitral valve. There is evidence of mitral regurgitation. There is also a slight thickening of the aortic valve. The ventricular functions are normal.

Infective endocarditis is an infection of the endocardial surface of the heart valves. It can involve previously normal valves, a valve damaged as a result of prior endocarditis, rheumatic heart disease or a degenerative process, or a prosthetic valve. Fibrin and platelets attach to the surface of damaged heart valves producing a sterile thrombotic endocarditis. Certain bacteria can adhere to these lesions on the valvular surface and initiate an inflammatory response that leads to further fibrin deposition, thus leading to maturation of the lesions. Such masses of fibrin and bacteria are termed 'vegetations'.

Apart from the cardiac findings of a new or changing murmur, patients with endocarditis can develop a number of other signs. Skin changes of infective endocarditis are common and include petechiae or purpura (blood spots under the skin) and splinter hemorrhages (linear reddish-brown lesions under the nail bed). Figure 2.1 shows an example of petechiae associated with endocarditis.

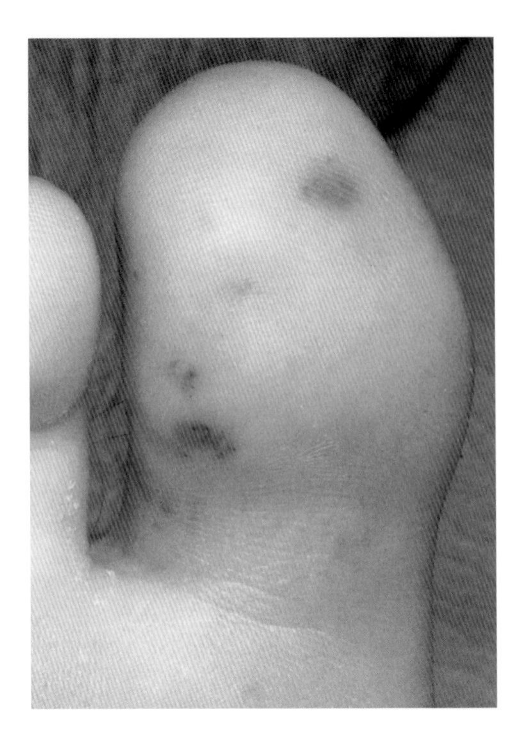

Figure 2.1 A petechial rash on the great toe of a patient with staphylococcal endocarditis. The rash appeared on the day after admission and was the first clue to the diagnosis.

Establishing the aetiology

Endocarditis leads to a sustained bacteraemia and this is best established by taking repeat cultures (three or more) with the first and the last cultures at least 1 hour apart.

You correctly took three sets of blood cultures. The microbiology laboratory phones to tell you that all three sets have now grown α-haemolytic streptococcus belonging to the viridans group. All isolates are sensitive to penicillin, erythromycin, clindamycin, vancomycin and tetracycline. Detailed testing subsequently identifies all isolates as Streptococcus sanguis.

Do you think this is a case of native mitral valve endocarditis?

It is indeed. On the ward round, your registrar mentions that you have satisfied two of the major criteria that are used to define endocarditis – sustained bacteraemia with a typical organism and an echocardiogram that is consistent with endocarditis.

Which criteria are being referred to?

The Duke criteria have been developed to aid the diagnosis of endocarditis – note that they were developed at Duke University in North Carolina, USA and there is no eponymous cardiologist! These criteria are useful in the clinical decision-making process if the diagnosis is difficult, e.g. fastidious bacteria may not grow on culture, prior antibiotics can inhibit bacterial growth, and often the echocardiographic abnormalities are too subtle to establish the presence of vegetations. The criteria are widely used in research on endocarditis to standardise patient selection.

How will you treat this patient?

You should seek expert opinion for the treatment of endocarditis and there are a variety of national treatment guidelines that you can consult. General principles are that bactericidal rather than static drugs should be used, and that extended therapy of weeks rather than days is required. A commonly used regimen for treatment of streptococcal endocarditis of a native heart valve is 4 weeks of benzylpenicillin combined with gentamicin for the first 2 weeks. Recent UK guidelines suggest that 2 weeks of treatment with benzylpenicillin and gentamicin may be enough if the penicillin minimum inhibitory concentration (MIC) for the organism is low (≤0.1mg/L) (see Antimicrobial chemotherapy, Part 1). Although streptococci are intrinsically resistant to gentamicin, benzylpenicillin disrupts the cell wall and allows gentamicin to penetrate inside the cell. Thus, they are synergistic when used in combination. The nature and duration of treatment depend upon the organism isolated and also the nature of the valve – native or prosthetic – and thus it is important to refer to the appropriate guideline.

What else do you need to do?

Do not forget to monitor the patient's general condition and blood results as several things could go wrong over the 4 weeks of treatment. In particular, you should monitor renal function as gentamicin is a nephrotoxic drug. Regular peak and trough levels of gentamicin need to be ascertained. Adequate peak levels imply that the patient is receiving enough gentamicin to produce a synergistic effect with penicillin. A high trough level correlates with increased toxicity. Speak to the ward pharmacists if in any doubt. Single-day dosing of gentamicin has become common in serious infection, but note that this has not been validated for the treatment of endocarditis. Remember that failure of medical treatment in endocarditis may precipitate the need for surgical intervention.

Your patient received 4 weeks of benzylpenicillin with gentamicin added for the first 2 weeks. According to the guidelines (see Further reading), there may have been a case for a shortened 2-week course, but the penicillin MIC of his Streptococcus sanguis isolate was 0.15mg/L and after discussion with the microbiologist you opted for the longer course. The patient's white cell count gradually came down and his CRP reverted to normal on day 18, but treatment was continued for 4 weeks. Your patient is cured and is due to go home.

What advice would you like to give him?

Patients with a history of endocarditis sometimes need antibiotic prophylaxis prior to certain procedures, particularly dental work. However, a review of published evidence by the National Institute for Health and Clinical Excellence (NICE) has considerably restricted the recommended indications for prophylaxis and this advice should be consulted when necessary.

CASE REVIEW

A 73-year-old man presented with a history of general malaise and a fever over a 2-week period. A pansystolic murmur detected on examination raised suspicions of endocarditis. This was confirmed by the presence of vegetations on the posterior leaflet of the mitral valve on echocardiograpy and by the isolation of *S. sanguis* from three separate sets of blood cultures. Four weeks of intravenous treatment with benzylpenicillin and gentamicin (latter for 2 weeks) was successful in curing the infection.

KEY POINTS

- The combination of a non-specific illness, a heart murmur and fever should always raise the suspicion of endocarditis.
- In suspected endocarditis, take several sets of blood cultures before commencing antibiotics.
- Treatment should be based on current guidelines and may require expert consultation. Long-term therapy (weeks) with bactericidal antibiotics is generally required.

Further reading and information

Durack DT, Lukes AS, Bright DK. New criteria for diagnosis of infective endocarditis: utilization of specific echocardiographic findings. Duke Endocarditis Service. *Am J Med* 1994; **96**(3): 200–9.

Elliott TSJ, Fowraker J, Gould FK, Perry JD, Sandoe JAT. Guidelines for the antibiotic treatment of endocarditis in adults: report of the Working Party of the British Society for Antimicrobial Chemotherapy. *J Antimicrob Chemother* 2004; **54**: 971–81.

http://www.medcalc.com/endocarditis.html.

http://www.nice.org.uk/guidance/index.jsp?action=download&o=40014.

Case 3 A 36-year-old man with hospital-acquired pneumonia

A 36-year-old man is admitted to the surgical ward with a 2-day history of vomiting and a sudden onset of severe abdominal pain. He is a habitual drinker and consumed excess alcohol a day prior to the onset of pain. He is afebrile on admission, his blood pressure is 116/78 mmHg and pulse 110 beats/min. Examination reveals marked tenderness in the epigastrium and the right subcostal region. Bowel sounds are normal and there is no mass felt on palpation.

Preliminary investigations reveal:

		(Normal range)
White blood cell count	*22 × 10^9/L*	*(4–11)*
Serum C-reactive protein	*256 mg/L*	*(<10)*
Serum alkaline phosphatase	*178 U/L*	*(45–105)*
Serum gamma glutamyl transferase	*437 U/L*	*(<50)*
Serum aspartate aminotransferase	*297 U/L*	*(1–31)*
Serum total bilirubin	*10 μmol/L*	*(1–22)*
Serum amylase	*2018 U/L*	*(60–180)*
Serum urea	*6.7 mmol/L*	*(2.5–7.5)*
Serum creatinine	*98 μmol/L*	*(60–110)*

What is the most likely diagnosis?

Acute pancreatitis is the single most likely cause of his admission. Predisposing factors for the development of acute pancreatitis include gallstones, alcohol intake, drugs such as thiazides, and viral infections.

He was treated conservatively with intravenous fluids and analgesics. However, the day after admission, he became hypotensive (BP 74/50 mmHg) and hypoxic and was transferred to the intensive care unit (ICU) for further management. An urgent computerised tomography (CT) scan of the abdomen revealed pancreatic necrosis. In the ICU he was put on a ventilator, given inotropic support and started empirically on IV meropenem 1 g tds.

His condition initially stabilised, but after 1 week in the ICU, he developed a fever and both clinical and radiological signs of pneumonia with cavitation in the right upper lobe.

What would you do now?

A bronchoalveolar lavage (BAL) is a very useful method of sampling the lower respiratory tract. A tracheal aspirate is easier to collect from a ventilated patient, but it samples the bacteriological flora that colonises the endotracheal tube and this does not necessarily reflect infection in the lower respiratory tract. The intensive care team sent a BAL for microbiological investigation requesting urgent microscopy and culture.

The microscopy revealed numerous Gram-positive cocci, scanty Gram-negative bacilli and numerous polymorphonuclear cells. Next day, the culture plates showed a profuse growth of Staphylococcus aureus, reported as >10^5 colony forming units of S. aureus/ml.

What does all this mean?

The patient has been in hospital for almost 10 days. New consolidation on the chest film, numerous polymorphonuclear cells on microscopy from a sample of BAL, and the profuse growth of *S. aureus* are all consistent with ventilator-acquired pneumonia. *S. aureus*, like so many other bacteria, can colonise the respiratory tract, particularly in hospitalised individuals. When isolated from sputum, *S. aureus* most commonly reflects nasopharyngeal colonisation, often following antibiotic treatment of a chest infection with agents that have no anti-staphylococcal activity. Thus *S. aureus* is a respira-

Infectious Disease: Clinical Cases Uncovered. By H. McKenzie, R. Laing, A. Mackenzie, P. Molyneaux and A. Bal. Published 2009 by Blackwell Publishing. ISBN 978-1-4051-6891-5.

tory pathogen only in certain circumstances, e.g. in patients with cystic fibrosis and as a complication of influenza. However, it can be a cause of ventilator-associated pneumonia, and a quantitative colony count on BAL cultures that demonstrates greater than 10^4 organisms/ml is good evidence of significant infection. You definitely want to know the antibiotic susceptibility of this organism.

What antibiotics do you commonly use for treating *S. aureus* infection?

Most *S. aureus* strains (>80%) are now resistant to benzylpenicillin by virtue of β-lactamase, an enzyme which breaks down penicillin. These can be treated by β-lactamase-stable penicillins (e.g. flucloxacillin) or by agents from other antibiotic groups (e.g. clindamycin, erythromycin and tetracyclines). However, a growing number of infections are caused by *S. aureus* strains that have developed a different mechanism of resistance and are resistant to β-lactamase-stable penicillins (see Antimicrobial chemotherapy, Part 1). These are called methicillin-resistant *S. aureus* or MRSA. Methicilllin is not used clinically because of its toxicity, but can be used to represent flucloxacillin in laboratory testing.

The laboratory informs you that it will take one more day before it can issue a sensitivity report. Whilst awaiting the report, the intensive care consultant decides to start IV vancomycin 1 g bd.

The next day, the microbiologist on the ICU round confirms that the strain is indeed an MRSA. It is susceptible to vancomycin, rifampicin, trimethoprim-sulphamethoxazole, linezolid and gentamicin. After some discussion, the patient was switched to linezolid.

Is the patient on appropriate treatment? What else do you need to do?

Vancomycin is an appropriate choice for MRSA pneumonia, while teicoplanin, another glycopeptide, is also an option. There is evidence that the new antibiotic linezolid is at least as good as vancomycin for MRSA pneumonia and there is some debate on which is the best option at present. Remember that vancomycin is nephrotoxic and you have to monitor levels. However, linezolid also has side effects – it causes marrow toxicity and therefore blood counts should be monitored and extended therapy beyond 28 days avoided. Treatment options for MRSA are limited and linezolid is one of the few oral agents available. Daptomycin is another new agent with activity against MRSA, but its optimum role in treatment is still under investigation. Daptomycin is not indicated in the treatment of chest infections.

How did the patient get infected with MRSA?

MRSA infections are often acquired in the hospital. MRSA is transferred from one patient to another via the hands of healthcare workers. Frequent handwashing is one of the best ways to stop the spread of MRSA. Is there any other patient in the ICU who has got MRSA infection or colonisation? Infection control staff should be informed, although it is more than likely that they will already be aware of it as they will keep all culture results for ICU patients under close surveillance. They will investigate the likely source of the infection and try to ensure there is no further spread.

Broad spectrum antibiotics such as the third generation cephalosporins and meropenem select MRSA because it is intrinsically resistant to these agents. Other bacteria are inhibited or killed, leaving an 'ecological space' into which MRSA can expand.

Note that there is an important difference between colonisation and infection (see Infection and immunity, Part 1). A patient, or indeed a member of staff or visitor to the ICU, may be colonised by MRSA (e.g. in the throat, axillae, perineum, etc.) but remain in perfectly good health. In the current case, MRSA was cultured from an invasive procedure which enabled the lower respiratory tract to be sampled. The lower respiratory tract should normally be sterile, so this culture result, taken together with the fever and right basal consolidation, provided strong evidence that the patient had an MRSA pneumonia. Isolation of MRSA from a throat swab or tracheal aspirate in this patient would not have been nearly so conclusive, as the nasopharynx or the endotracheal tube could easily be colonised with something other than the cause of the pneumonia.

What else would you do now?

The patient should be isolated in a single room and needs to be barrier nursed (see Table N, Approach to the patient, Part 1). You must wear disposable gloves and apron when examining a patient who is colonised or infected with MRSA. Always wash your hands or apply an alcohol hand rub after examining a patient who is colonised or infected with MRSA – in fact you should do so after every patient contact.

What happened next?

The patient was treated with linezolid for 2 weeks and made a slow recovery. He was transferred to general surgery ward after 3 weeks in the ICU.

What steps will you take when the patient is being transferred?

The ward doctors and nurses need to know that the patient is positive for MRSA. The patient needs to be isolated and barrier nursed even after leaving the ICU. Advice from microbiologists and/or your hospital's infection control team should be sought and they would have a plan in place depending upon local circumstances and conventions.

Are there any other organisms that need specific infection control measures to prevent their spread?

Yes, and the list is ever increasing. Next to MRSA, *Clostridium difficile*, which causes antibiotic-associated diarrhoea (CDAD), is probably the commonest infection control hazard. Like MRSA, it is preferentially selected out by broad spectrum antibiotics. This is a spore-forming organism and the spores are particularly difficult to get rid of from the environment.

It is also important to control the spread of bacteria that are resistant to multiple antibiotics. Patients who carry vancomycin-resistant enterococci (VREs) or coliform organisms with extended spectrum β-lactamases (ESBLs) should also be the subject of infection control measures. Other organisms that present a risk of spreading from patient to patient include varicella zoster virus (VZV), norovirus, group A streptococci, tuberculosis and respiratory viruses such as respiratory syncytial virus (RSV) and influenza.

CASE REVIEW

A 36-year-old man was admitted with vomiting and severe abdominal pain, quickly diagnosed as acute pancreatitis. He was transferred to the ICU on the day following his admission as he became hypoxic and his blood pressure dropped to 74/50 mmHg. His condition initially stabilised but unfortunately after a week in the ICU he developed ventilator-associated pneumonia caused by MRSA. This was treated with linezolid and he made a slow recovery, spending 3 weeks in the ICU.

KEY POINTS

- Healthcare-acquired infection extends hospital stays and prevention is better than cure.
- Ventilator-associated pneumonia is common in intensive care units and bronchoalveolar lavage gives the best opportunity for accurate microbiological diagnosis.
- There are relatively few treatment options for MRSA infection, but some new agents are emerging.

Further reading and information

American Thoracic Society and the Infectious Diseases Society of America. ATS/IDSA Guidelines: guidelines for the management of adults with HAP, VAP and HCAP. *Am J Respir Crit Care Med* 2005; **171**: 388.

Gemmell CG, Edwards DI, Fraise AP *et al*. Guidelines for the prophylaxis and treatment of methicillin resistant *Staphylococcus aureus* (MRSA) infections in the UK. *J Antimicrob Chemother* 2006; **57**: 589–608.

Masterton RG, Galloway A, French G *et al*. Guidelines for the management of hospital acquired pneumonia in the UK: report of the Working Party on Hospital-Acquired Pneumonia of the British Society for Antimicrobial Chemotherapy. *J Antimicrob Chemother* 2008; **62**: 5–34.

Case 4 A 26-year-old male with jaundice

A 26-year-old male presents to his general practitioner complaining of malaise, loss of appetite and mild abdominal pain. He has noticed that his urine is darker than usual. He looks mildly jaundiced.

What are the possible causes of jaundice?

Jaundice can be pre-hepatic, hepatic and post-hepatic. The general practitioner sent some baseline bloods to the biochemistry laboratory. The results were available on computer the following day:

		(Normal range)
Serum total bilirubin	*78 μmol/L*	*(1–22)*
Serum alanine aminotransferase	*1780 U/L*	*(5–35)*
Serum alkaline phosphatase	*198 U/L*	*(45–105)*
Serum urea	*6.5 mmol/L*	*(2.5–7.5)*
Serum sodium	*138 mmol/L*	*(137–144)*
Serum potassium	*5.4 mmol/L*	*(3.5–4.9)*
Haemoglobin	*14.5 g/dL*	*(13–18)*
White blood cell count	*8.6 × 10^9/L*	*(4–11)*
Platelets	*145 × 10^9/L*	*(150–400)*
Prothrombin time	*15 s*	*(11.5–15.5)*

On the basis of these results, the GP refers the patient to the infectious disease unit with a provisional diagnosis of viral hepatitis.

What causes of hepatitis should be considered?

1 Hepatitis A, B, C or E.
2 Epstein–Barr virus.
3 Cytomegalovirus.
4 Alcohol.
5 Autoimmune.
6 Drug related.
7 Other infections that are less common causes: leptospirosis, Q fever, Lyme disease, brucellosis, syphilis.

Infectious Disease: Clinical Cases Uncovered. By H. McKenzie, R. Laing, A. Mackenzie, P. Molyneaux and A. Bal. Published 2009 by Blackwell Publishing. ISBN 978-1-4051-6891-5.

What details would you like to elicit from the history?

Presenting complaint and history of presenting complaint

- Duration of each symptom.
- Location of abdominal pain.

Associated symptoms

- Myalgia.
- Nausea or vomiting.
- Fever.
- Pruritis.
- Rash.
- Arthralgia.
- Enlarged glands or sore throat.
- Any family member or contact unwell or jaundiced.

Past medical history/drug, family and social history

- Occupation.
- Injecting drug use.
- Drug history: including prescribed, over the counter and recreational.
- Previous immunisation against hepatitis A or B.
- Previous jaundice.
- Alcohol use.
- Foreign travel.
- Involvement in watersports.
- Sexual orientation.
- Sexual exposures.

The history reveals that he has been unwell for about 10 days, initially with malaise, anorexia, nausea and myalgia and feeling as if he might be getting flu. Although the anorexia and general flu-like feeling are not too bad, the malaise continues, he now has a persistent ache in his right upper quadrant and his urine is much darker than usual. He has had no itch, rash, joint symptoms, enlarged glands or sore throat. He lives with his partner, currently 4 months pregnant, and one child aged 4 years. He has an office job and has not been outside Europe in the previous year. The last time he was abroad was a family holiday to Lanzarote 4 months previously. He does not recall being immunised against hepatitis, knows of no-one with jaundice and does not remember ever being jaundiced. He is on no prescribed medication and drinks little alcohol. He is bisexual and has had one male sexual contact over the last 6 months. He admits to recreational cannabis but denies injecting drug use.

Review your differential diagnoses

1 *Hepatitis A or B*: plausible in view of the short illness. Hepatitis E is also a possibility, but less likely in someone who has not travelled outside Europe.
2 *Epstein–Barr virus*: possible, but no enlarged glands or sore throat. More common in teenagers.
3 *Cytomegalovirus*: possible, but subclinical or mild illness is most common.
4 *Alcohol*: unlikely if history is accurate.
5 *Autoimmune*: more common in females.
6 *Drug-related*: unlikely.
7 *Other infections that can cause hepatitis*: leptospirosis, Q fever, Lyme, brucellosis, syphilis – no obvious risk factors for these other than perhaps the bisexual history (syphilis).

What virology investigations should be done?

Hepatitis A IgM, hepatitis B surface antigen and hepatitis C antibody. Hepatitis E serology is only available in specialist laboratories and should be considered if all other tests are negative.

The following virology results are telephoned the next day:

Hepatitis A IgM	*Positive*
Hepatitis B surface antigen	*Positive*
Hepatitis B core IgM antibody	*Negative*
Hepatitis B core total antibody	*Positive*
Hepatitis B e-antigen	*Negative*
Hepatitis B e-antibody	*Positive*
Hepatitis C antibody	*Negative*

How do you interpret these tests?

The results confirm recent hepatitis A in a patient with chronic hepatitis B infection. A positive IgM is diagnostic of recent or current infection with the relevant pathogen, so the positive hepatitis A IgM result explains the current presentation.

The positive hepatitis B surface antigen shows that the patient is also infected with hepatitis B. The absence of hepatitis B core IgM means that it is not an acute hepatitis B infection – there is core antibody (total is positive) but it must be IgG, reflecting infection some time ago. Hepatitis e-antigen is a marker of high infectivity if positive.

Note that if the patient had been immunised, or had cleared his hepatitis B infection, he would have been negative for surface antigen and might be positive for surface antibody. It is easy to become confused when interpreting serological results – especially in hepatitis B – so you should seek advice from a virologist if you are unsure. A guide to the interpretation of hepatitis tests is given in the Table 4.1, but you might also find it useful to review serological diagnosis in Part 1 (Laboratory diagnosis of infection).

What further investigations should be considered and why?

You should discuss with the patient the possibility of testing for human immunodeficiency virus (HIV) and syphilis as both these infections are more common in homosexual and bisexual men.

Following appropriate counselling, these tests are carried out and the results are both negative.

Do these results exclude infection with HIV and hepatitis C?

No, as both tests measure the antibody response, which can be delayed. HIV antibody may take up to 3 months and hepatitis C antibody up to 6 months to become positive after exposure.

What should be done about the diagnoses of acute hepatitis A and chronic hepatitis B?

Hepatitis A

Symptomatic management is indicated, with hospital referral of the patient if necessary. Local health protection

Table 4.1 Guide to hepatitis tests.

Clinical scenario	Test required	Test result	Answer to question	Comment
Is this illness due to hepatitis A?	Hepatitis A IgM	Positive	Yes	A negative result makes hepatitis A unlikely as this IgM is usually detectable at the onset of illness
Does my patient have hepatitis B?	Hepatitis B surface antigen	Positive	Yes	Manage patient as infectious
Is this non-immunised patient susceptible to hepatitis B?	Hepatitis B anti-core antibody	Negative	Yes	This antibody is not induced by immunisation but does indicate previous or current disease. When the anti-core antibody is positive, other tests are needed to interpret the significance as it can be positive in acute infection, chronic infection or previous infection There is also a hepatitis B IgM anti-core antibody test. When this IgM test is positive, it is consistent with recent hepatitis B, but low positives are not uncommon in chronic infection
Is my patient protected from hepatitis B infection?	Hepatitis B anti-surface antibody	Positive	Yes	This antibody is produced following immunisation or natural infection
Does my patient have active hepatitis C infection?	Hepatitis C RNA	Positive	Yes	Except following a known exposure incident or in an immunocompromised patient, RNA for hepatitis C is usually tested for only if hepatitis C antibody is positive. It can take up to 6 months for antibody to become positive and it may not develop in the immunocompromised

staff should be informed (it is a notifiable infection), to enable them to identify contacts and possible sources of the infection – remember it is usually transferred by the faecal–oral route.

Contacts

Active (killed vaccine) and passive (human normal immunoglobulin, HNIG) immunisation are available for protection of contacts. HNIG is usually restricted to selected close household contacts.

Hepatitis B

The patient should be referred (non-urgently) to a specialist with an interest in liver disease for further investigation and management of his chronic hepatitis B.

Contacts

Testing and immunisation of household and sexual contacts should be offered in accordance with the advice of the Department of Health. Again the local health protection department should be notified of this case.

The patient remained only mildly unwell and made an uneventful recovery from hepatitis A infection. However, note that it is important to follow up the pregnant partner and child. If the female partner has contracted hepatitis B, then the fetus is at risk of vertical transmission. If not, and she has not been previously immunised, that should be done now. The 4-year-old child should also be tested. Testing and interpretation of results in this kind of situation is complex and you may need to take specialist advice from a virologist.

What immunisations are available in viral hepatitis?

Passive immunisation is fast acting but short lived, and involves the transfer of pre-existing antibody. No immunological memory is created. HNIG is produced from pooled blood donations and has enough hepatitis A antibody for it to be useful as prophylaxis in susceptible patients following potential exposure to the disease. Specific hepatitis B immunoglobulin (HBIG) prepared from blood donors with high levels of immunity is available to give similar protection in those exposed to hepatitis B infection.

Active vaccine takes longer to give protection, but is long lasting and creates immunological memory. An antigenic preparation prepared from the virus is administered and stimulates protective immunity. In hepatitis A the vaccine is an inactivated preparation of the virus, while in hepatitis B the vaccine consists of surface antigen produced by yeast cells in culture after the relevant gene has been transferred by recombinant DNA technology.

There are no available vaccines for hepatitis C or E. Hepatitis E (clinically similar to hepatitis A) is usually transmitted by the faecal–oral route so good standards of food and water hygiene are the mainstay of prevention.

Blood-borne viruses in healthcare workers

All healthcare workers who start to perform exposure prone procedures (EPPs) in a new role are now screened for the presence of blood-borne virus infections prior to employment to ensure that their patients are not put at risk. The definition of an EPP is based on the risk that injury to the worker may result in exposure of the patient's open tissues to the blood of the worker. This includes those invasive procedures where the worker's gloved hands may be in contact with sharp instruments, needle tips or sharp tissues (spicules of bone or teeth) inside a patient's open body cavity, wound or confined anatomical space, where the hands or fingertips may not be completely visible at all times. Simple venepuncture or putting up drips are examples of procedures that are not considered to be EPPs.

CASE REVIEW

A 26-year-old male presented with malaise, loss of appetite and mild abdominal pain. His urine was darker than usual and he looked mildly jaundiced. Serological testing revealed acute hepatitis A infection against a background of chronic hepatitis B infection. The treatment of hepatitis A infection is symptomatic but he will be referred for specialist management of his chronic hepatitis B. His pregnant partner and 4-year-old child require screening and immunisation if appropriate.

KEY POINTS

- Hepatitis viral serology can seem complex and you may need expert help in interpreting it.
- IgM positive results usually reflect recent or current infection.
- Acute hepatitis A infection is generally managed symptomatically.
- Hepatitis A and B are both notifiable infections.
- Passive and active vaccination is available for both hepatitis A and B and contacts should be screened and offered immunisation as appropriate by the health protection department.
- Healthcare staff who perform exposure prone procedures should be screened for blood-borne viruses according to national Department of Health guidelines.

Further reading and information

British Liver Trust information on hepatitis A and B is available at www.britishlivertrust.org.uk/home.aspx.

Department of Health guidance on immunisation in the UK is available at http://www.dh.gov.uk/.

Department of Health advice on health clearance of new healthcare workers for tuberculosis, hepatitis B, hepatitis C and HIV and more information on exposure prone procedures is also available at http://www.dh.gov.uk/.

Case 5 A 28-year-old female student with severe headache and drowsiness

A 28-year-old student was taken to the accident and emergency department by ambulance with a 1-day history of severe headache, neck pain, fever and drowsiness. She is accompanied by her friend.

What differential diagnosis would you think of at this stage?

1 Bacterial meningitis.
2 Viral meningitis.
3 Subarachnoid haemorrhage (SAH).
4 Migraine.
5 Sinusitis.
6 Pneumonia.
7 Urinary tract infection.
8 Influenza.
9 Occult bacteraemia.

What details would you like to elicit from the history?

History of presenting complaint

Find out more about the headache:

- Duration: hours, days or months?
- Onset: gradual or sudden?
- Location: frontal, lateral, occipital or global?
- Previous similar symptoms.
- Any other family members or contacts affected?

Associated symptoms

- Nausea: very non-specific; may be present in infection and other conditions.
- Vomiting: non-specific, but suggests significant illness.
- Photophobia: part of 'meningism' and suggestive of meningitis.
- Rash: may have been noticed by patient – 'purple spots that do not go away under the glass test'.
- Rigors: very suggestive of bacteraemia.
- Flashing lights: suggestive of migraine.

Note that the terms meningism and meningitis have different meanings (Box 5.1).

Past medical history

- Migraine: usually a unilateral frontal headache with nausea and visual disturbance – compare with current symptoms.
- Hypertension: subacute cause of headache and a risk factor for SAH.
- Sinusitis: a cause of meningism or a risk for meningitis.
- Middle ear infections: a risk for meningitis.
- Head trauma or surgery: a risk for meningitis.
- Splenectomy: a risk for overwhelming sepsis.
- Immunocompromised: HIV, haematological malignancy.

Drug history

- Antimicrobials – recent or current.
- Immunosuppressives: steroids, chemotherapy, cyclosporin, cyclophosphamide, azathioprine.
- Oral contraceptive – may cause headache.

Allergies

- Penicillins or cephalosporins may cause true allergy (e.g. rash), anaphylaxis or non-allergic adverse reaction (e.g. vomiting).
- It is important to clarify this as many life-threatening infections are treated with β-lactam antibiotics.

Family history

- Migraine: may run in families.
- Meningitis: recent family member may be affected.
- Subarachnoid haemorrhage: may run in families.

Infectious Disease: Clinical Cases Uncovered. By H. McKenzie, R. Laing, A. Mackenzie, P. Molyneaux and A. Bal. Published 2009 by Blackwell Publishing. ISBN 978-1-4051-6891-5.

Box 5.1 Definitions of meningism and meningitis

Meningism	Triad of headache, neck stiffness and photophobia	A frequent clinical observation in a variety of conditions, e.g. pharyngitis, sinusitis, pyelonephritis
Meningitis	Inflammation of the meninges, demonstrated by a raised white cell count in the cerebrospinal fluid (CSF)	Usually due to infection, but occasionally due to other causes, e.g. malignancy or sarcoidosis

Social history

- Travel history:
  - Mecca or sub-Saharan Africa: meningococcal meningitis.
  - Spain: penicillin-resistant pneumococci.
  - Tropics: cryptococcal meningitis.
- Heavy alcohol: a risk for listeria meningitis.

Further history from her and her friend reveals that the headache was of gradual onset, and associated with nausea, vomiting and photophobia. She has a history of migraine headaches but this 'feels different'. Recent sore throat and runny nose, then felt 'blocked up'. There is no history of recent antibiotic therapy. She returned from holiday in Spain 13 days ago.

Review your major differential diagnoses

1 *Bacterial meningitis*. The triad of headache, photophobia and neck stiffness ('meningism') suggests meningitis. This may have either a bacterial or viral aetiology, but other causes of meningism are frequently encountered. The drowsiness is a cause for concern and points to bacterial rather than viral meningitis. Being relatively young and a student, one thinks of meningococcal (*Neisseria meningitidis*) meningitis but she is older than the typical age range (16–24 years) for this condition. The absence of rash (according to the patient) does not rule out meningococcal meningitis as the rash reflects meningococcaemia, which may not be present, and the rash can evolve so quickly the patient may not have noticed it.

Her preceding upper respiratory tract symptoms can occur in meningococcal meningitis but are more suggestive of pneumococcal (*Streptococcus pneumoniae*) meningitis. The significance of the travel history is that she may have contracted a penicillin-resistant strain of pneumococcus in Spain, where penicillin-resistance rates are high.

Other causes are less likely because of her age and less common, but still possible. *Haemophilus influenzae* more commonly caused meningitis in children, but has become unusual since the introduction of the Hib (*H. influenzae* type b) conjugate vaccine. *Listeria monocytogenes* occurs in neonates and in middle-aged and elderly patients with co-morbidity such as diabetes and alcoholism.

2 *Viral meningitis*. This is more common than bacterial meningitis. It is most commonly due to an enterovirus, e.g. echovirus, coxsackie. Headache may be severe, but conscious level is generally normal. Viral meningitis is seasonal and commoner in the summer and autumn months.

Other causes of meningism include sinusitis, otitis media, pharyngitis, tonsillitis, pneumonia and pyelonephritis.

3 *SAH*. Often occurs in older patients. Typically presents with a sudden onset of occipital headache, 'like being struck on the back of the head with a bat'. May have meningism, reduced conscious level and low-grade fever – so it can be similar to meningitis.

4 *Migraine*. Severe migraine can be difficult to distinguish from meningitis in a patient with a history of migraine. If there is no history of migraine it is unlikely to present like this and further investigation is mandatory.

You now plan to examine the patient. What are the key things to look for?

General examination

- Rash: a typical purpuric (purple, non-blanching) rash suggests meningococcal infection; this condition can also present with an unusual transient, pink, blotchy rash (which blanches on pressure) in its subacute form. A purpuric rash may also occasionally occur with pneumococcal infection, due to disseminated intravascular coagulation (DIC).
- Reduced conscious level: this is a vital diagnostic and prognostic sign. The Glasgow Coma Scale (GCS) is gen-

erally used to assess consciousness. Note that if if the GCS is <8 the patient will probably be unable to protect her airway, and intubation and ventilation should be considered.

- Fever (>38 °C): this is suggestive of infection (or inflammation). A low temperature (<35.5 °C) suggests the same. Normal temperature measurements on admission do not rule out serious infection.
- Tachycardia (pulse >100 beats/min): part of systemic inflammatory response syndrome (SIRS).
- Hypotension (systolic <100 mmHg, diastolic <60 mmHg): this is a vital sign. This may be due to hypovolaemia (dehydration), but may also be due to septic shock (see Infection and immunity, Part 1), in which case the patient will need inotropic support (e.g. with noradrenaline) in an intensive care unit.
- Ear discharge: this suggests a middle ear infection (otitis media), with possible perforation of the ear drum. Generally due to *S. pneumoniae*, *H. influenzae* or, if chronic, *Pseudomonas aeruginosa*.
- Pharyngitis/tonsillitis: adequate examination with a tongue depressor and a torch may reveal an obvious diagnosis to explain the meningism.

Chest examination

Tachypnoea (respiratory rate >20 breaths/min), reduced oxygen saturation (<95%) on pulse oximetry, and signs of consolidation are suggestive of pneumonia.

Cardiovascular examination

Endocarditis tends to present insidiously with fever and malaise rather than with meningism, but look for a cardiac murmur and other stigmata.

Abdominal examination

Pyelonephritis is a common cause of meningism and may present with fever and loin tenderness, with relatively few urinary symptoms.

Central nervous system examination

- Look for a reduced conscious level – GCS score.
- Feel for neck stiffness – patient must be lying flat in bed without pillow. The neck may be very stiff and the back may be arched backwards (opisthotonus).
- Examine for focal neurological signs – hemiparesis, cerebellar signs, visual field defect and dysphasia all suggest a focal lesion in the brain (e.g. abscess or infarct). It may be difficult for the patient to cooperate with the examination because of confusion, which itself is a more global deficit.
- Look for papilloedema – it may not be possible to view the fundi because of photophobia. If present, papilloedema suggests raised intracranial pressure (ICP) but it is not a reliable sign as ICP may be significantly raised in the absence of papilloedema. Raised ICP is an important development because it may lead to potentially fatal cerebellar herniation ('coning') at the time of lumbar puncture. A CT scan may give further indications of a raised ICP.

On examination the patient appeared drowsy (GCS 10: eyes open to pain, localises pain, confused speech) with no rash. She had a temperature of 39.5 °C, pulse 120 beats/min and BP 95/55 mmHg. The pharynx, chest, cardiovascular system (CVS) and abdomen were normal. The neck was stiff and there were no focal neurological signs. Fundi were normal. There was tenderness over the right mastoid sinus.

What is your working diagnosis now?

Bacterial meningitis is the most likely of the differential diagnoses. It is *not* possible to distinguish between meningococcal and pneumococcal meningitis at this stage. The absence of rash makes meningococcal meningitis less likely. The tenderness over the right mastoid sinus raises the possibility of mastoiditis, with contiguous spread to the meninges.

What would you do now?

If you can delay antibiotic treatment for long enough to enable you to collect a set of blood cultures, this will increase your chances of getting a microbiological diagnosis. However, this is a matter of clinical judgement and depends on the severity of the patient's condition. Patients with suspected meningococcal sepsis should have been given antibiotic by the attending GP before admission to hospital – in acute meningococcal sepsis there is no time to be lost in treatment.

A set of blood cultures were taken, a full blood count sent to haematology and blood for urea and creatinine and liver function tests sent to biochemistry. Immediately after these samples had been taken, the patient was given IV ceftriaxone 2 g, and started on 0.15 mg/kg dexamethasone four times a day. An urgent CT head scan and chest X-ray were requested.

Empirical antibiotic therapy

Traditionally, bacterial meningitis presenting in this way was treated with intravenous high-dose benzylpenicillin. More recently, therapy with an intravenous third generation cephalosporin (e.g. ceftriaxone or cefotaxime) has been recommended (British Infection Society (BIS) guidelines) because of the increasing prevalence of penicillin resistance among pneumococci and, to a lesser extent, penicillin resistance among meningococci. The cephalosporin will also treat other less likely pathogens, e.g. *Haemophilus influenzae*, but not *Listeria monocytogenes*, so if this is suspected, high-dose intravenous amoxicillin should be added. The BIS guidelines suggest adding amoxicillin empirically in individuals over 55 years of age. Note that treatment can be rationalised once an organism has been identified and its antimicrobial susceptibility tested.

Empirical steroids

The role of steroids (e.g. IV dexamethasone) in the treatment of bacterial meningitis was debated for many years, a consensus view being hindered by a relative lack of clinical studies. The current view, extrapolated from some studies showing reduced mortality and fewer complications in those with severe infection treated with steroids, is that steroid should be given to patients with suspected bacterial meningitis with altered conscious level, seizures or other evidence of raised intracranial pressure.

Bloods

Blood cultures may be positive, allowing a microbiological diagnosis to be made, even if antibiotic has previously been given (most commercially available blood culture bottles contain an antibiotic-binding resin). Blood can be tested by the polymerase chain reaction (PCR) for the meningococcus or pneumococcus. However, this is usually available only in reference laboratories and you should be guided by your local laboratory. Take blood for a full blood count, coagulation screen, CRP, glucose, urea and electrolytes and liver function tests.

- A full blood count may show an elevated (or sometimes low) white cell count suggestive of infection. The platelet count and haemoglobin may be reduced, the coagulation prolonged and the blood film can show red cell fragmentation if DIC is present. DIC is a feature of severe sepsis and carries a grave prognosis.
- The CRP is typically elevated in bacterial infection, but may be normal or only slightly elevated on admission as it lags behind the clinical condition of the patient by about 24 hours.
- The serum glucose is important as hypoglycaemia may cause confusion or reduced conscious level: hyperglycaemia often occurs as a response to infection, and could be a predisposing factor. The CSF glucose measurement obtained at lumbar puncture must be interpreted with reference to the serum glucose.
- The serum sodium is commonly low due to a mixture of dehydration and the syndrome of inappropriate antidiuretic hormone secretion (SIADH) due to sepsis. The urea and creatinine may be elevated due to pre-renal renal dysfunction as a result of dehydration and hypotension, or due to acute tubular necrosis in severe sepsis.
- The albumin is likely to be reduced in acute infection and the liver enzymes may be mildly abnormal in any infection. If the liver enzymes are moderately or grossly elevated, this suggests a systemic infection or inflammatory process.

Imaging

- A CT head scan is *not* routinely required in suspected meningitis. Indications would be altered conscious level including confusion, seizures and focal neurological signs, all of which raise the possibility of elevated intracranial pressure.
- A scan may also be performed if another diagnosis, e.g. SAH, cerebral haemorrhage or brain abscess, is thought more likely than meningitis.
- Chest X-ray: this should be routine, as meningitis may be associated with pneumonia (e.g. pneumococcal and rarely *Mycoplasma pneumoniae*). Meningitis may be complicated by adult respiratory distress syndrome (ARDS), which causes respiratory failure and is likely to require ventilatory support.

> *You get a telephone report of the CT head as 'normal, except for complete opacification of right mastoid sinus, and some opacification of frontal and ethmoid sinuses'. You therefore proceed to lumbar puncture (LP).*

Lumbar puncture

This is the diagnostic test of choice and the only way to make a definitive diagnosis of meningitis. It should be performed on all patients with suspected bacterial or viral meningitis, unless there is a contraindication such as features of raised intracerebral pressure (see comments on coning above). However, if the typical purpuric rash of meningococcal infection is present, LP is generally not performed because the rash is almost pathognomonic and the LP would delay therapy.

The opening pressure of the CSF should be measured using a manometer: it is normally 11–19 cmH_2O. The CSF pressure is typically elevated (>30 cmH_2O) in bacterial meningitis, sometimes spilling over the top of the manometer tube in severe infection. The CSF, which is normally clear (often described as gin clear) will probably be cloudy, or even appear as frank pus, in meningococcal or pneumococcal meningitis.

Normal CSF should not have any white blood cells (WBCs) (<5 cells/µL permitted) or red blood cells (RBCs), although some may be present due to a traumatic LP ('bloody tap'). The presence of several thousand WBCs/µL, mainly polymorphs, is typical of a bacterial meningitis. In viral meningitis there are typically up to 1000 WBCs/µL, mostly lymphocytes.

The CSF protein is typically >1.0 g/L in bacterial meningitis and only slightly elevated (0.7–1.0 g/L) in viral disease. CSF glucose is typically about two-thirds of the serum glucose (samples should be collected at the same time for comparison) and this is unchanged in viral meningitis, but CSF glucose is markedly reduced (<50%) in bacterial meningitis. The CSF glucose can be an important result in considering the likely diagnosis if previous antibiotic therapy means that a bacterial aetiology is not immediately evident.

A Gram stain of the CSF may show organisms, e.g. Gram-negative diplococci (*Neisseria meningitidis*) or Gram-positive diplococci (*Streptococcus pneumoniae*). Whilst positive findings are diagnostic, negative findings do not rule out bacterial meningitis. Antigen tests for the common pathogens that cause bacterial meningitis are routinely performed, but while their good specificity(>90%) allows confidence in a positive test, their relative poor sensitivity (~60%) means that a negative test does not rule out infection due to that pathogen. PCR may be performed for the meningococcus or pneumococcus or other specific organisms if other methods fail to identify the pathogen.

The following results are back – how do you interpret them?

- *Initial blood tests*

		(Normal range)
Haemoglobin	*15.5 g/dL*	*(11.5–16.5)*
White blood cell count	*18.0 × 10^9/L*	*(4–11)*
Platelet count	*334 × 10^9/L*	*(150–400)*
Serum C-reactive protein	*150 mg/L*	*(<10)*
Serum glucose	*6.9 mmol/L*	*(3–6 for fasting plasma glucose)*
Serum sodium	*133 mmol/L*	*(137–144)*
Serum potassium	*3.9 mmol/L*	*(3.5–4.9)*
Serum urea	*12.2 mmol/L*	*(2.5–7.5)*
Serum creatinine	*111 µmol/L*	*(60–110)*
Serum albumin	*33 g/L*	*(37–49)*
Serum alanine aminotransferase	*41 U/L*	*(5–35)*
Serum alkaline phosphatase	*110 U/L*	*(45–105)*
Serum gamma glutamyl transferase	*35 U/L*	*(4–35)*

- *Chest X-ray: normal*
- *Lumbar puncture:*

CSF opening pressure	*34 cmH_2O*	*(5–18)*
CSF appearance	*Cloudy*	
CSF total protein	*3.21 g/L*	*(0.15–0.45)*
CSF glucose	*1.1 mmol/L*	*(3.3–4.4)*
CSF white blood cell count	*3200/µL (95% polymorphs)*	
CSF red blood cell count	*0*	
CSF Gram stain	*No organisms seen*	
CSF antigen tests	*Negative*	
CSF culture	*Awaited*	

What is your diagnosis now?

This is almost certainly bacterial meningitis because the high white cell count is predominantly polymorphs (would be predominantly lymphocytes in viral meningitis) and the CSF glucose is around 16% of the serum value (1.1/6.9), rather than 66% as you would expect. However, the Gram stain and antigen tests were negative so it is still not possible to distinguish between meningococcal and pneumococcal meningitis.

Twelve hours later she is more lucid and generally better. The microbiology registrar telephones to say that the blood cultures are positive in the automated blood culture machine and microscopy reveals Gram-positive diplococci in both bottles.

What does this mean?

Firstly, the patient is bacteraemic and presumably this is the organism that has caused the meningitis. The presence of bacteraemia conveys a worse prognosis and

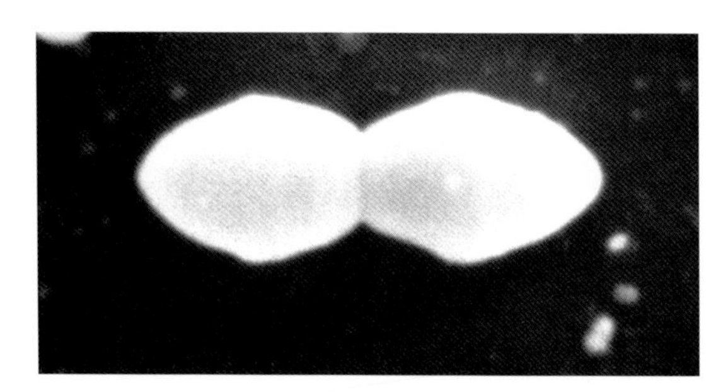

Figure 5.1 Scanning electron micrograph of pneumococcal diplococci.

the patient is at increased risk of severe sepsis or septic shock.

The Gram-positive diplococcus is likely to be *S. pneumoniae*, which is a common cause of meningitis in a patient of this age. Pneumococci often appear to be 'paired' in a Gram stain and that characteristic appearance raises the suspicion of pneumococcal infection (Fig. 5.1). An antigen test on the blood culture at this stage can help confirm that the organism is a pneumococcus, but identity is not absolutely confirmed until the organism grows on culture. It is interesting that no organism has been identified in the CSF at this stage but this is frequently the case. Pre-hospital antibiotic treatment by the GP is recommended and this can sometimes result in no growth.

Rates of penicillin resistance among pneumococci in the UK are generally low (2–5%) but may be up to 60% in some countries such as Spain, South Africa and Hungary. The patient has recently been to Spain but she is already on intravenous ceftriaxone to which it is likely to be susceptible, unless it is fully resistant to penicillin. The results of the sensitivity testing should be available tomorrow, and if fully sensitive, consideration can be given to changing her treatment to benzylpenicillin (narrower spectrum).

The opacification in the sinuses, especially the right mastoid, is the likely source of the meningitis and the bacteraemia. The mastoid abnormality raises the possibility of an underlying anatomical problem such as a cholesteatoma or a skull fracture, and a detailed CT scan, or a magnetic resonance imaging (MRI) scan of this area is indicated to look for these abnormalities.

An MRI scan of the right mastoid and others sinuses show no underlying problem, and the following day she is afebrile, normotensive and lucid.

In addition to treatment with antibiotic and steroid, what are the other important aspects of management?

- Because a patient such as this requires intensive nursing, monitoring and therapy, they are best treated in a high dependency unit or intensive care unit.
- Hyperglycaemia should be controlled with intravenous insulin if necessary.
- Adequate oxygenation should be maintained with supplemental oxygen, delivered via a mask, or by intubation and ventilation if necessary, most commonly because of unconsciousness.
- Hypotension should be corrected with intravenous fluid with reference to the central venous pressure (CVP), and an intravenous inotropic agent (e.g. adrenaline, noradrenaline) utilised if necessary to maintain adequate blood pressure.
- Early consideration should be given to artificial feeding via a nasogastric tube if the patient is not able to eat after 24–48 hours.

How long should the intravenous dexamethasone continue?

There is no evidence base for this decision but the consensus view is about 4 days. During this time, it is necessary to monitor for signs of hyperglycaemia, hypertension and confusion, all potential side effects of dexamethasone.

How long should the intravenous ceftriaxone continue?

The consensus view is that meningococcal meningitis should be treated for at least 7 days, but there is some evidence that shorter courses are effective. Pneumococcal meningitis should be treated for at least 10–14 days, as should haemophilus meningitis, and listeria meningitis for 21 days. These durations are for guidance only and the approach to treatment should be individualised for each patient depending on circumstances. There is the option to 'switch' to an appropriate oral agent (e.g. amoxicillin) to complete the course of treatment when the patient is afebrile and eating normally.

After 48 hours, the laboratory reported the growth of S. pneumoniae, fully sensitive to benzylpenicillin from both bottles of the blood culture. You decide to switch from ceftriaxone to benzylpenicillin for a further 8 days, since the narrower spectrum of activity makes the development of complications such as Clostridium difficile less likely.

What is the prognosis for meningitis?

- *Pneumococcal*: the mortality is about 10%. Up to 30% of surviving patients may suffer some long-term sequelae, e.g. deafness, epilepsy, cognitive impairment or headaches.
- *Meningococcal*: the mortality in meningitis is 10–50%. About 20% may have long-term sequelae. Note that the mortality in meningococcal sepsis without meningitis is even higher.

Should the relevant health protection department be notified about a case of meningitis?

All suspected or confirmed cases of meningococcal infection should be notified to the Health Protection Agency since the risk of further cases is significantly higher amongst contacts than in the rest of population. All close (sometimes described as 'kissing') contacts (e.g. family, room-mates, school contacts), and the index case before discharge from hospital, should receive prophylaxis with rifampicin or ciprofloxacin to eradicate the organism from the nasopharynx. Targeted immunisation may be used in outbreaks if the infecting strain is a group C meningococcus. Pneumococcal meningitis is not notifiable as there is no indication for prophylaxis.

Is meningitis preventable by vaccine?

There are several antigenically distinct groups of *Neisseria meningitidis* and groups B and C are the commonest in the UK. A conjugate vaccine that gives good protection against group C strains (MenC) is now part of routine childhood immunisation in the UK. It is recommended that all individuals under 25 years should be immunised and this is particularly important for any unprotected individual attending university, since outbreaks amongst 'freshers' occur from time to time. The majority of the remaining infections in the UK are now group B, for which there is not yet an effective vaccine. Group A strains are responsible for epidemics in sub-Saharan Africa, and there have been outbreaks of groups A and W135 infection associated with the annual Hajj pilgrimage to Mecca in Saudi Arabia. Immunisation with the quadrivalent A, C, W135 and Y polysaccharide vaccine (not a conjugate vaccine) is now an entry requirement for Saudi Arabia and this has helped to control cases.

There are over 90 different capsular serotypes of *S. pneumoniae* and there is no generic vaccine that covers all possible strains. Two types of vaccine are currently used. Pneumococcal polysaccharide vaccine (Pneumovax) includes 23 of the most common serotypes and is used in the UK in adults over 65 years and younger adults in high-risk groups. A conjugate vaccine (Prevenar) containing seven common serotypes is part of the routine childhood immunisation schedule in the UK.

The Hib conjugate vaccine is part of the routine childhood immunisation schedule in the UK and this has resulted in a dramatic reduction in the incidence of meningitis due to *Haemophilus influenzae*. There is no vaccine against *Listeria monocytogenes*.

CASE REVIEW

A 28-year-old female student presented with severe headache, fever and drowsiness. The patient was treated on admission as a matter of urgency with ceftriaxone and steroids. In the absence of any contraindications, a lumbar puncture was performed and the high CSF polymorph count and low glucose strongly supported a diagnosis of bacterial meningitis. Antigen tests and Gram stain on the CSF were negative but after 12 hours a blood culture became positive on microscopy with Gram-positive diplococci, later confirmed as *S. pneumoniae*, sensitive to benzylpenicillin. The patient was switched to benzylpenicillin and made a full recovery after 10 days of treatment.

KEY POINTS

- Classic features of meningitis are headache, photophobia and neck stiffness.
- Always consider meningitis in a patient with altered consciousness, especially if accompanied by fever.
- Treat any patient with suspected meningococcal infection immediately with parenteral benzylpenicillin or ceftriaxone.
- Blood cultures are often helpful in diagnosis, especially if a lumbar puncture is contraindicated.

Further reading and information

Heyderman RS *et al.* Early management of suspected meningitis and meningococcal septicaemia in immunocompetent adults. *J Infect* 2003; **46**: 75–7.

The British Infection Society guidelines on management are available on the Meningitis Research Foundation website: http://www.meningitis.org.

Case 6 A 30-year-old female with a vesicular skin rash

A 30-year-old female in-patient mentions she has noticed a rash. She shows you some vesicles on one arm.

What details would you like to elicit from the history?

Presenting complaint and history of presenting complaint

- Onset: was it acute or chronic, what is the duration?
- Distribution: is it on the arms, legs, face, etc?
- Type of lesions and how they develop or change.
- Previous similar rash?
- Contact with anyone with similar rash?
- Why is she an in-patient?

Associated symptoms

- Fever.
- Itch or other abnormal skin sensation.

Past medical history/family, social and drug history

- Any immune deficiency?
- Could she be pregnant?
- Occupation.
- Current and recent drug therapy.

What common terms are used to describe a rash?

Table 6.1 lists some common terms for describing rashes.

Your patient tells you she noticed a rash that morning when washing, present only on the one arm, which is a little itchy and which she has never previously had. As well as the vesicles, there are some reddish lesions. Nothing was noted when she was admitted yesterday for investigations because of deteriorating renal function. She had a renal transplant 5 years previously and is taking prednisolone and cyclosporin. There has been no fever and she has been generally well recently. She has had no contact with any rashes. She is not pregnant and works part-time at a school nursery. Your examination confirms that the rash is localised to the left C6 dermatome on the forearm. There are vesicles of varying size, some with surrounding erythema and some papules.

What infections do you consider in the differential diagnosis for a vesicular rash?

1 Varicella zoster virus (VZV): shingles if localised, chickenpox or disseminated shingles if widespread, or vaccine related if recently immunised with the live attenuated vaccine.
2 Herpes simplex virus (HSV).
3 Enteroviruses: hand, foot and mouth disease causes vesicles which can be present on any or all of the palms, soles, buccal mucosa, buttocks and genitals.

Because the rash is vesicular, you suspect shingles or possibly herpes simplex – what should be done?

Source isolation of the patient (see Approach to the patient, Part 1) is necessary as the patient presents a risk of infection to other patients, especially the immunocompromised, and to pregnant staff.

You take a swab from the base of a vesicle for virus detection. You are phoned later that day to say the swab is positive for VZV antigen by immunofluorescence (see Laboratory diagnosis of infection, Part 1), confirming the diagnosis as shingles.

What should you do now for the following?

- The patient.
- Other patients: she has been in a four-bedded sideward since admission.
- Staff: at least one ward staff member is pregnant.

Infectious Disease: Clinical Cases Uncovered. By H. McKenzie, R. Laing, A. Mackenzie, P. Molyneaux and A. Bal. Published 2009 by Blackwell Publishing. ISBN 978-1-4051-6891-5.

Table 6.1 Some common terms for describing rashes.

Rash description	Appearance	Common causes or associations
Macular	The spots are well circumscribed but not raised – lesions cannot be felt by running a finger over the skin	Parvovirus B19, rubella
Maculopapular	Mixture of flat spots (macules) and raised lesions (papules) detectable by palpation	Measles Erythema multiforme (may be a feature of *Mycoplasma* infection)
Purpuric	Red or purple spots caused by bleeding beneath the skin – do not blanche on pressure ('tumbler test')	Meningococcaemia
Vesicular	Fluid-filled, circumscribed, raised lesions less than 0.5 cm across	Chickenpox or shingles (VZV), herpes simplex, hand, foot and mouth disease
Nodular	Thickened palpable lesions that are deep to the skin surface	Erythema nodosum (may be post-streptococcal)

The patient

Prescribe aciclovir (appropriate dose for renal function). Either keep the patient isolated until all lesions are crusted or discharge her home with plans to readmit when no longer infectious.

What needs to be established regarding other patients and staff?

• Who has had significant contact with this patient while she is infectious?
• Are they susceptible to chickenpox?

Why is it necessary to do this?

There are infection control implications. Passive immunisation with varicella zoster immunoglobulin (VZIG) may be indicated for those who have had significant contact and are susceptible to chickenpox. This is especially important for:
• Neonates.
• Pregnant women.
• Immunocompromised patients.

What are the possible complications for the neonate, pregnant or immunocompromised should they acquire chickenpox?

• *Pregnant women:* a pregnant woman may be at risk of severe chickenpox and complications, especially varicella pneumonia.
• *Fetuses:* up to 20 weeks' gestation, the fetus is at risk of fetal varicella syndrome. Although rare, affecting only 1–2% of pregnancies with chickenpox at this gestation, adverse outcome includes neurological abnormalities, limb hypoplasia and skin scarring with high mortality. If infection is acquired by the mother within 1 week of delivery, the neonate is at risk of life-threatening neonatal varicella because of transplacental infection. Following intrauterine VZV infection, shingles may occur at an unusually early age in a healthy child.
• *Neonates:* if exposed within the first 7 days of life, fulminant chickenpox may occur, which has a high mortality and significant morbidity in the survivors.
• *Immunocompromised patients:* life-threatening complications, e.g. pneumonitis.

What should be done?

• Who has had significant contact? Ask the nurse in charge to establish which patients, in addition to those in the four-bedded side ward, the patient may have had significant contact with and also which staff may have had significant contact.
• Of these significant contacts, who is susceptible to chickenpox?
• At this stage you should seek help from the hospital's infection control team in managing the process.

What information is needed to establish if a VZV contact has been significant?

A significant contact is dependent on:

• *Contact during time of infectivity:* shingles is infectious from the development of vesicles until the lesions are all crusted; chickenpox is infectious from 48 hours prior to the rash until the lesions are all crusted.

• *Location of rash:* during the time of infectivity, exposed shingles lesions are infectious and any chickenpox lesions are infectious.

• *Underlying illness:* for shingles in the immunocompromised, unexposed lesions should also be considered infectious.

• *Closeness and duration of contact:* any face-to-face contact or 15 minutes in the same house, classroom or four-bedded ward.

The criteria for what constitutes a significant contact with both chickenpox and shingles are available in the Green Book (see Further reading).

If a susceptible person has significant exposure to chickenpox or shingles, they should be considered potentially infectious from 8 to 21 days after exposure.

> *The advice of the infection control nurse is that as the rash has sometimes been exposed and as the patient is immunocompromised, any person who has had face-to-face contact or been in her side ward for 15 minutes today should be considered a significant contact. The rash was not present when she was 'clerked in' on the day of admission, so only today's contacts are being identified.*

How do you decide susceptibility to chickenpox for exposed patients or staff who have had significant contact?

Ask about previous chickenpox or shingles. Except for the immunocompromised, a previous history of definite chickenpox or shingles means they are not susceptible. Of all the significant patient and staff contacts, the following might be susceptible:

• The three patients in the same side ward: all are immunocompromised in some way and one is uncertain about previous chickenpox.

• A nurse who is 12 weeks' pregnant and a visiting male doctor from abroad, who was on the morning ward round, are both fairly sure they have not had chickenpox.

What action should be taken now?

Blood should be taken from the three patients and the two healthcare workers and tested for VZV IgG as soon as possible. Phone to tell the laboratory samples are coming and ask when the results should be available.

> *The results are available the next day. All three patients are VZV IgG positive. You reassure them that they are not at risk of catching chickenpox. Both the nurse and the doctor are VZV IgG negative.*

What should be done now for the susceptible pregnant nurse?

Passive immunisation with VZIG should be advised as soon as possible. Although unnecessary delay should be discouraged, the administration of VZIG is indicated for up to 10 days for a pregnant contact (up to 7 days after contact for an immunocompromised contact).

The nurse receives the VZIG – what else needs to be considered?

Infection control issues in relation to her work.

For how long should she be considered potentially infectious?

The period of potential infectivity is extended after VZIG. She should be considered infectious from 8 to 28 days after exposure and excluded from contact with those at increased risk of severe chickenpox during this time.

What should be done about the doctor?

The susceptible doctor is due to leave and return home to India in 5 days. He is not immunocompromised. As he would be potentially infectious from 8 to 21 days after the contact, he does not need to be excluded from patient contact for the rest of his visit. He should be told when he should consider himself infectious. In the meantime he contacted the hospital occupational health service to discuss his options:

• Immunisation with live, attenuated VZV vaccine.

• No immediate action, but to be prepared to start aciclovir immediately if a rash appears (ideally started within 24 hours).

Are there any other common causes of a vesicular rash?

The typical vesicular rash of chickenpox is shown in Plate 9, but there are other causes of vesicular eruptions. Hand, foot and mouth disease is an enteroviral infection and relatively common. The main route of spread is faecal–oral so good hand hygiene prevents transmission. The causal enteroviruses in the UK are not associated with problems in pregnancy and pregnant women who come into contact with cases or develop lesions can be reassured. In general, the diagnosis is clinical, but where laboratory confirmation is wanted, virus detection (now usually by PCR) can be done on faeces or a lesion swab.

CASE REVIEW

A 30-year-old female renal transplant patient was admitted to a four-bedded side ward for worsening renal function. The day after admission, she was noted to have shingles on her left forearm (C6). She had been looked after by a pregnant nurse who was negative for varicella IgG and therefore at risk of acquiring chickenpox. The nurse was given VZIG. Three other patients who shared the ward were found to be positive for varicella IgG and therefore not at risk of infection. A visiting doctor who was a significant contact and susceptible was found not to be in any of the risk groups that require VZIG.

KEY POINTS

- Shingles can transmit VZV to susceptible individuals and cause chickenpox.
- Chickenpox can be a very severe illness in adults (especially pregnant women), neonates and the immunocompromised.
- It is important to determine whether contact with a case of chickenpox or shingles is significant or not. Expert opinion should be sought (e.g. communication with virologist or infection control) if in doubt.
- Active and passive immunisation for VZV are available.
- Aciclovir is ideally started within 24 hours of the rash appearing.
- Hand, foot and mouth disease is an important differential diagnosis which has no implications in pregnancy.

Further reading and information

Department of Health advice on VZV immunisation is available in chapter 34 in the *Immunisation against Infectious Disease 2006* (the 'Green Book') available at http://www.dh.gov.uk/en/Policyandguidance/Healthandsocialcaretopics/Greenbook/DH_4097254. This chapter also defines a significant VZV contact.

Royal College of Obstetricians and Gynaecologists advice on chickenpox in pregnancy, *Green-top Guideline No. 13*, is available at www.rcog.org.uk.

Tunbridge AJ, Breuer J, Jeffery KJM, on behalf of the British Infection Society. Chickenpox in adults – clinical management. *J Infect* 2008; **57**: 95–102.

Case 7 A 37-year-old man with fever and pleuritic pain

A 37-year-old farmer was admitted to hospital with a 3-day history of fevers, rigors and right-sided chest pain.

Suggest some possible causes for this man's symptoms

1 Pneumonia.
2 Pulmonary embolism.
3 Costochondritis/rib osteomyelitis.
4 Pericarditis.
5 Lung abscess.
6 Skin/soft tissue infection of the chest wall.

What details would you like to elicit from the history?

Presenting complaint and history of presenting complaint

- Find out more about the pain:
  - Localisation – is it just right-sided or does it extend or radiate elsewhere?
  - Severity and whether there has been any relief from analgesics.
  - Previous similar symptoms.
  - Onset – abrupt or gradual?
  - Aggravating factors, e.g. breathing or movement.
- Find out more about the other symptoms:
  - Does he have rigors or shivers? It is worth clarifying this since rigors, if present, are suggestive of bacteraemia.
  - Does he have a cough? If so, is the cough dry or productive and has he coughed up any blood (haemoptysis)?
  - Is he breathless? This may be due to difficulty taking a deep breath as a result of pleuritic pain or it may be due to disease involving the lung leading to impaired gas exchange.

Other symptoms

- Diarrhoea or vomiting – these can be prominent symptoms with some pneumonias (especially Legionnaire's disease).
- Leg swelling – a unilateral leg swelling would arouse suspicion of deep venous thrombosis leading to pulmonary embolism.

Past medical history/drug history

- Has this patient ever had a similar illness? A recurrent pneumonia would raise concerns about an underlying immune deficiency (e.g. HIV disease) or lung disease (e.g. bronchogenic carcinoma or bronchiectasis).
- Is there a past history of venous thrombosis or pulmonary embolism?
- Is he on any medication? He may have been started on antibiotics from his family doctor before hospital admission and this could influence the results of your investigations (e.g. sputum cultures may be negative).
- Has he been in hospital in the last 6 weeks? If an infection has been picked up in hospital it may be resistant to many of the more commonly used antibiotics.

Family history/social history

- Occupation – what sort of farm does he have? Livestock may be a source of unusual causes of pneumonia such as *Coxiella burnetii* (Q fever).
- Travel history – this is very important when considering a patient with possible pneumonia since imported respiratory pathogens may need to be treated with different antibiotics from those usually used in the UK (e.g. imported pneumococcal pneumonias are more likely to be resistant to penicillin).
- Does he smoke? This could be important in terms of increased risk of both respiratory infection and venous thrombosis.

Infectious Disease: Clinical Cases Uncovered. By H. McKenzie, R. Laing, A. Mackenzie, P. Molyneaux and A. Bal. Published 2009 by Blackwell Publishing. ISBN 978-1-4051-6891-5.

Further history reveals that the patient had previously been fit and healthy. His right-sided chest pain was severe, sharp and worse on inspiration and coughing. There was no rib tenderness. He gave a clear history of uncontrolled violent shaking (rigors) which had been so dramatic that his wife had thought he was having a seizure. He had a dry cough but no haemoptysis. He was a smoker but had no past history of chest problems. He had worked on a mixed farm (cattle, sheep and arable) for the last 15 years and had not travelled abroad since a trip to Florida 6 months earlier. He had received no medication from his GP prior to admission and had not been in hospital since his tonsillectomy at the age of 15 years.

Time to review your differential diagnosis

1 *Pneumonia.* This would seem the most likely diagnosis. Do not be deterred by the lack of sputum production as this can often lag behind the symptoms of fever and pleuritic chest pain. Even if he was not a smoker, pneumonia would be the most likely diagnosis.
2 *Pulmonary embolism.* It can be very difficult to differentiate between a pulmonary infarct and pneumonia. Patients with both conditions may present with pleuritic chest pain and fever. Both may give a history of haemoptysis. In this case, this history of rigors would point more towards infection and away from infarction.
3 *Costochondritis/rib osteomyelitis.* The pain of costochondritis (inflammation of the rib cartilage usually caused by a viral infection) can be very severe and similar to pleuritic pain but it would not usually be associated with rigors. Furthermore you would expect localised tenderness to be present in patients with costochondritis or rib osteomyelitis.
4 *Pericarditis.* Pain and fever are typically present in pericarditis. The pain will, like pleurisy, usually be worse on inspiration. In contrast to pleurisy, pericarditic pain tends to be central (retrosternal) and may be relieved on sitting forwards.
5 *Lung abscess.* A lung abscess is less likely and would generally present with a longer history of illness and less prominent pleuritic chest pain.
6 *Skin/soft tissue infection of the chest wall.* This should become obvious on examination but the possibility of a local infection giving rise to pain and fever needs to be borne in mind at this stage of the history. Localised tenderness would be a feature of superficial infection. Remember also that some patients will present with severe chest wall pain as the first symptom of shingles – even before the rash has appeared!

What next?

Examination – what are the key things to look for here? You should be asking yourself:
• Do the findings on clinical examination support my top differential diagnosis of pneumonia?
• If this patient has pneumonia, what is there in the clinical examination to help me decide on the severity of the illness? This is especially important as the admitting doctor needs to decide as early as possible whether the patient can be safely managed in a general medical ward or whether a high dependency unit (HDU)/ICU would be a more appropriate setting.

On examination the patient appeared breathless and was struggling to answer questions. When he did speak, his answers were often confused or incomplete. His respiratory rate was measured as 35 breaths/min. He had a temperature of 38.5 °C, pulse 115 beats/min and BP 105/55 mmHg. On examining the back of the chest, there was a dull percussion note over the right lower lobe. Auscultation revealed bronchial breathing over the lower right chest.

What is the likely diagnosis?

Community-acquired pneumonia, i.e. one that has developed in a patient who has not been in a hospital or other healthcare institution.

What would you do now?

Tests or treatment?

Do not get bogged down thinking about which tests to do in a case like this, think of the treatment that needs to be initiated immediately.

Oxygen

The patient is obviously breathless with signs of consolidation on chest examination so he needs oxygen. Give high-flow oxygen, e.g. 15 L/min through a rebreathing mask.

Fluids

Do not forget the importance of fluid resuscitation in any patient with clinically severe infection. Patients with pneumonia may not only be vasodilated as a result of the reaction to infection but will also be fluid depleted through insensible losses (fever and tachypnoea), so fluids are a vital component of immediate treatment.

Antibiotics

The patient is obviously too unwell to be given oral antibiotics so these need to be administered intravenously. Before starting treatment, remember to take blood cultures. A positive blood culture can help you to direct your antibiotic treatment effectively when sensitivity tests become available. The recommended antibiotic for community-acquired pneumonia will vary from hospital to hospital but it is important to recognise that clinical features, blood biochemistry, blood haematology and chest X-ray appearances are not reliable ways of differentiating one type of pneumonia from another. Initial antibiotic treatment should therefore be broad spectrum.

In severe community-acquired pneumonia, the British Thoracic Society guidelines recommend a combination of a β-lactamase-stable β-lactam together with a macrolide (e.g. co-amoxiclav plus clarithromycin). The β-lactam covers *Streptococcus pneumoniae* (the commonest cause) and the macrolide covers the so-called 'atypical' causes of pneumonia, e.g. *Mycoplasma pneumoniae*, legionnaire's disease, Q fever and psittacosis. The term 'atypical' is widely used for pneumonia caused by these organisms and they are atypical in the sense that none of them can be identified by simple overnight culture of sputum and a range of other tests is needed. However, in clinical terms, 'atypical' pneumonias do not present as a clearly recognisable entity that can be distinguished from 'typical' pneumonia, so the term is now discouraged.

The patient was given immediate doses of co-amoxiclav and clarithromycin intravenously.

After initiating these treatments, consider the appropriate tests.

Bloods

As well as taking blood cultures you need to take venous blood for urea and electrolytes and a full blood count. Take arterial blood for blood gas analysis.

Imaging

A chest X-ray should be requested. Note that in some cases, the patient may be too ill to attend the X-ray department and a portable chest X-ray may need to be arranged on the ward. Figure 7.1 shows the patient's chest X-ray.

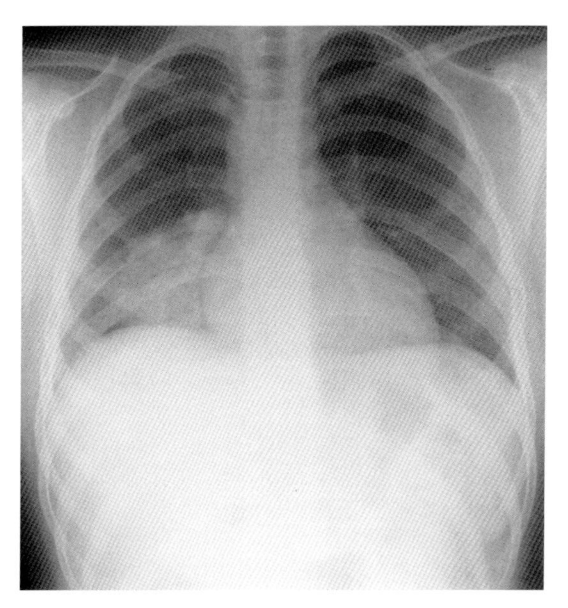

Figure 7.1 Chest X-ray showing consolidation of the right lower zone. Courtesy of Professor F. J. Gilbert.

Cultures

As well as blood, sputum should be sent for microscopy and culture if it can be obtained.

Serology

Take a baseline (acute) blood sample for serological tests and put a reminder in the notes to take a follow-up (convalescent) sample in 7–10 days (see Serology in Laboratory diagnosis of infection, Part 1). Such tests can be useful in the diagnosis of infection with 'atypical' organisms (see above), but will not give you an answer for several days and may not always be positive early on in the infection.

Antigen tests

Urine should be sent for antigen tests. Tests for *S. pneumoniae* and *Legionella pneumophila* antigens are available and the latter is particularly useful since the organism is not easily or rapidly cultured. However, negative tests do not exclude these infections

Time to review the patient

Thirty minutes after starting treatment with fluids, oxygen and antibiotics the patient was re-examined. The pulse was

now 110 beats/min and regular, BP 110/60 mmHg, respiratory rate 35 breaths/min and temperature 38 °C. He appeared less responsive than earlier. Some test results have returned as follows:

		(Normal range)
Haemoglobin	*13.7 g/dL*	*(13–18)*
White blood cell count	*21.5 × 10^9/L*	*(4–11)*
Platelet count	*170 × 10^9/L*	*(150–400)*
Serum sodium	*130 mmol/L*	*(137–144)*
Serum potassium	*3.2 mmol/L*	*(3.5–4.9)*
Serum urea	*8.2 mmol/L*	*(2.5–7.5)*

Blood gas analysis breathing 15 L oxygen/min:

PO_2	*7.5 kPa*	*(11.3–12.6)*
PCO_2	*5.8 kPa*	*(4.7–6)*
pH	*7.05*	*(7.36–7.44)*
Base excess	*−5 mmol/L*	*(±2)*

The high white cell count would be consistent with a pneumonia but does not in itself tell you about the type or severity of infection. The worrying results are:

- Raised serum urea.
- Low PO_2 despite high-flow oxygen.
- Acidosis and negative base excess.

Raised urea is a poor prognostic marker in patients with pneumonia and is one of a number of features that are used to determine CURB-65 score (Box 7.1).

This patient has four out of a possible five poor prognostic features, i.e. everything except age >65 years. This is associated with a high mortality (Box 7.2).

Patients with a CURB-65 score of 3 or more have severe community-acquired pneumonia and require urgent treatment. Patients with a CURB-65 score of 2 may need short-term hospital admission, whereas patients with a score of 0 or 1 can usually be managed at home.

The blood gas analysis shows that this patient has ventilatory failure and needs support. The acidosis reflects tissue hypoxia and sepsis.

What needs to be done?

This patient needed to be transferred to an ICU for a period of assisted ventilation, intensive monitoring and possibly other support – such as inotropes to improve his blood pressure and circulation.

Box 7.1 CURB-65 scoring system

New confusion
Urea >7
Tachypnoea (respiratory rate >30 breaths/min)
Hypotension (systolic BP <90 or diastolic BP <60 mmHg)
Age >65 years

Box 7.2 Mortality related to the CURB-65 score

No. of CURB-65 features	Mortality
0	0.7%
1	3.2%
2	13%
3	17%
4	41.5%
5	57%

What happened next?

The patient was transferred to ICU where he was ventilated for 4 days before being well enough to be taken off the ventilator and transferred back to the medical ward. On the day after admission the laboratory telephoned to say that Gram-positive diplococci were seen on the blood cultures. The organism was confirmed as S. pneumoniae that was sensitive to penicillin. The patient's antibiotic treatment was de-escalated to benzylpenicillin. He was discharged from hospital after 2 weeks and returned 4 weeks later for a repeat chest X-ray, which showed that the lung had returned to normal.

What else might have happened?

Even a patient with pneumonia who receives the correct antibiotic in a timely fashion can still develop complications. The commonest one to look out for is an empyema. This is essentially an abscess in the pleural space and is likely to be associated with a persistent fever despite antibiotics. Sometimes a patient can initially improve, with the fever resolving on therapy, only to develop further fevers and a rising white cell count several days later. This should always prompt you to think about the possibility of an empyema. In these circumstances the diagnosis can usually be confirmed by a chest X-ray which will show a pleural collection. Aspiration and examination of the pleural fluid will usually show a low pH (<7.2) and the presence of white cells and organisms. Treatment requires a chest drain and appropriate antibiotics, but even this may not be enough, in which case surgical removal of the empyema is needed.

CASE REVIEW

A 37-year-old man was admitted with symptoms consistent with pneumonia. On examination he was febrile, tachycardic and tachypnoeic. His white cell count was high, the blood urea was elevated and he was acidotic. He was administered oxygen, fluids and intravenous co-amoxiclav and clarithromycin and required a transfer to the intensive care unit where he was ventilated. After blood cultures taken at admission grew *S. pneumoniae*, sensitive to penicillin, he was switched to benzylpenicillin. He made a good recovery after 2 weeks in hospital.

KEY POINTS

- Patients presenting with severe community-acquired pneumonia need urgent intervention.
- Initial assessment should include eliciting a careful history, carrying out a thorough examination, and investigations such as chest X-ray, a full blood count, urea and electrolytes, and liver function tests and blood gases.
- Microbiology tests include sputum for culture, blood cultures, urine for antigen testing, and clotted blood as a baseline for serology (needs further specimen in 7–10 days time).
- Adhere to local or national guidelines in selecting antibiotic treatment. In severe pneumonia a common regimen would be a combination of a β-lactamase-stable β-lactam antibiotic with a macrolide, e.g. co-amoxiclav and clarithromycin.
- The CURB-65 score is a useful prognostic tool in patients with pneumonia and can inform the need for hospital admission and for intravenous rather than oral antibiotics.

Further reading and information

www.brit-thoracic.org.uk.
www.sign.ac.uk/guidelines/fulltext/59.

Case 8 A 14-year-old school girl with a rash

A 32-year-old mother attends your GP surgery with her 14-year-old daughter who had to be collected from school after the school phoned to say her daughter had developed a rash. You see a girl who does not look acutely ill, but from a distance looks as if she has a maculopapular rash.

What are some likely diagnoses?

1 Infection, especially viruses.
2 Allergy.
3 Drug reaction.

What details would you like to elicit from the history?

History of presenting complaint

Find out more about the rash:
- Nature of rash: macular, maculopapular, purpura, vesicles or nodules (see Case 6)?
- Distribution of rash: localised or generalised on limbs/face/torso bilateral or unilateral?
- Onset: how it started and how it has developed.
- Any other family members and/or contacts affected, e.g. at school?

Associated symptoms

Enquire about:
- Fever.
- Lymph glands (lymphadenopathy).
- Appetite, nausea, vomiting or bowel upset.
- Sore throat, cough or conjunctivitis.
- Current and recent general health.
- Itch.

Infectious Disease: Clinical Cases Uncovered. By H. McKenzie, R. Laing, A. Mackenzie, P. Molyneaux and A. Bal. Published 2009 by Blackwell Publishing. ISBN 978-1-4051-6891-5.

Past medical history

- Immunisation history.

Drug history

- Any recent prescribed or over-the-counter medication?

Allergies

- Any known allergies.

Social history

More possibilities may need consideration depending on:
- Travel history.
- Animal contact.
- Change in soap or clothes washing products used.

Further history from the mother and daughter reveals that the daughter had been feeling mildly unwell for a few days, but not serious enough to be off school. Both she and her mother thought she was improving. She is keen not to miss school because of imminent exams. When changing for gym she noticed a blotchy rash on her trunk, both upper arms and face. There is no definite history of fever or glands, no gastrointestinal symptoms other than slight decrease in appetite, because of some nausea, and no itch, cough or conjunctivitis. Her throat is slightly sore. She was not immunised with the measles, mumps and rubella (MMR) vaccination because of the previous adverse publicity, but all other immunisations are up to date. She denies taking any medicines, knows of no allergies, has not been away from home for several months and has no new contact with any animals. Your examination reveals nothing of note other than a widespread maculopapular rash and a few mildly tender cervical lymph nodes. There is no pharyngitis, conjunctivitis, abdominal tenderness or masses.

Review your differentiated diagnoses

1 *Viral infection:* this is highly probable.

2 *Allergy:* this is possible, but there is no obvious precipitant.

3 *Drug reaction:* this is excluded if the history is accurate.

What do you plan to advise and do for this presumed viral rash?

• *Discuss diagnosis.* Even with investigations, a precise diagnosis may not be made. However, it is quite possible this is one of the so-called rash illnesses of childhood. It could be due to parvovirus B19 which causes slapped cheek syndrome (also known as erythema infectiosum or fifth disease). This is a common infection and, being aware of other current cases in the community, you think this is the most likely cause. Rubella is less likely, despite her not having had the MMR, because most children have been immunised and it is not currently circulating in the community. However, the rash of rubella can be clinically indistinguishable from that of parvovirus B19. It does not look like measles, but there are other viral infections which could cause recent mild illness followed by a rash.

• *Discuss investigations.* A diagnosis of parvovirus B19 or rubella is possible by a blood test. Would the mother like this? Would the daughter agree to a blood test?

• *Discuss management.* Conservative management with reassurance and no treatment or investigations is appropriate for most well children with a suspected viral rash. If it is due to parvovirus B19, the rash may last a number of days and recur over 1–3 weeks, especially when she is hot (e.g. after exercise or a hot bath). Transient arthropathy may occur.

What is your advice about school attendance?

As she feels well and you are confident in your clinical diagnosis of parvovirus B19, she can attend school. Infectivity is prior to the development of rash, so school exclusion for a well child is inappropriate.

What about school attendance if you had been uncertain of your diagnosis?

With both measles and rubella, infectivity is maximal prior to the rash, but for both school exclusion is recommended for 5 days from the onset of the rash, even if the child is well, so as to prevent transmission.

What else might you do for a presumed viral rash?

If either measles or rubella is suspected, both are notifiable infections so the local health protection department should be informed promptly. This should be done on clinical suspicion without awaiting laboratory confirmation.

Why is this prompt notification important?

Measles is highly infectious and if immunisation levels in the community are not high many susceptible people may be infected quickly. Even where immunisation levels are high overall, outbreaks may develop in poorly immunised subpopulations of a community. The health protection departments can both identify and arrange appropriate management of close contacts for whom protective action may be necessary and also arrange salivary testing of the patient for measles and rubella (i.e. a non-invasive sample); this would usually be appropriate, other than during a large epidemic. Oral fluid can be used to detect both viral IgM and nucleic acid.

What protection is available for close contacts of measles?

MMR can protect against measles if given within 3 days of contact and HNIG offers protection if given within 6 days of contact. MMR contains live attenuated measles, mumps and rubella viruses, and contraindications include immunocompromised patients and pregnant women. HNIG can be used where MMR is contraindicated. Prior to giving HNIG, the patient's immune status (measles IgG) should be established, if there is time, in order to avoid giving it to those who already have immunity and therefore do not require it. Any opportunity should be taken to review an individual's immunisation history and those, like this child, who have not received MMR should be offered it unless there is a genuine contraindication. If over 10 years old, two doses of MMR at least 1 month apart are needed to induce long-lasting immunity.

Why are rashes important in pregnant women?

Several viruses that cause a rash illness can lead to complications in pregnancy (Box 8.1).

Box 8.1 Viruses that can cause complications in pregnancy

Rubella	Transplacental infection of the fetus can cause multiple problems including cataracts, deafness and cardiac defects
Parvovirus B19	Transplacental infection of the fetus can cause fetal anaemia and lead to fetal hydrops
Measles	Intrauterine death and pre-term delivery

As well as transplacental, how else can these viruses spread?

- Droplet: measles, rubella and parvovirus B19.
- Blood: parvovirus B19 because of the very high viraemia.

Who else may have serious illness with parvovirus B19 and why?

Virus replication occurs in red cell precursors in the bone marrow, causing cell death. In immunocompromised patients, infection becomes persistent as their immune response cannot control the viral replication. Red cell aplasia develops, their haemoglobin falls and chronic anaemia may develop.

In patients with shortened red cell survival time (e.g. in haemolytic disorders), an acute life-threatening anaemia – a transient aplastic crisis (TAC) – may develop due to the sudden temporary cessation of erythropoiesis.

The daughter agrees to have a blood test. You request investigation for parvovirus B19 and rubella. You give the mother and child current information regarding MMR including the Department of Health website, and ask them to read about it and either arrange with the practice nurse for immunisation or, if they wish to discuss further, to make a further appointment with you.

The mother says that she is not pregnant. However, a friend visiting their house recently is 16 weeks pregnant and a classmate at school had a renal transplant 18 months previously.

What do you advise and why?

Both these people should be told that parvovirus B19 infection is suspected. The pregnant woman should contact her midwife so that she can be investigated for parvovirus B19, both IgG and IgM, according to the Health Protection Agency guidance. About 50–60% of adults in the UK have had previous infection and are not at risk, but fetal anaemia and pregnancy loss are a risk at this stage of pregnancy and can occur without maternal symptoms. Obstetric investigations can detect the development of fetal anaemia for which intrauterine transfusion may be indicated as it improves survival.

The renal physician looking after the transplant recipient should also be contacted. As with the pregnant contact, testing for parvovirus B19 should be done.

After some days, the results you receive are:

Parvovirus B19	*IgM positive, IgG negative*
Rubella virus	*IgM negative, IgG negative*

What do these results mean and what do you do?

These results are consistent with recent infection with parvovirus B19 and no previous infection (or immunisation) with rubella. You let the mother and child know the diagnosis and ask them to pass this information on to the at-risk contacts.

Outcomes

The child makes an uneventful recovery. At a subsequent consultation with yourself it is agreed that the child will be immunised with MMR and this takes place uneventfully with no recognised sequelae.

The pregnant woman had a blood test and found she had been infected with parvovirus B19 before, even though she was unaware of this. The kidney transplant patient had several blood tests as he was susceptible. He remained well, IgG and IgM negative to parvovirus B19 with stable haemoglobin over the subsequent 4 weeks.

CASE REVIEW

A 14-year-old girl had a maculopapular rash. She was not too unwell. Tests revealed a positive parvovirus B19 IgM with a negative IgG consistent with a recent infection. Tests for rubella (both IgM and IgG) were negative indicating susceptibility. The GP took the opportunity to update her MMR immunisation status.

KEY POINTS

- In illnesses associated with a rash, it is important to ascertain the nature of the rash.
- Although a person with a rash may need no specific management, susceptible contacts may be at risk of serious illness.
- A maculopapular rash has several possible viral causes including rubella, parvovirus B19 and measles.
- Exposure to rashes in pregnancy and immunocompromised patients is a common concern and the 'Green Book' – available on line at the Department of Health website – is a useful source of advice.
- An attempt should be made to update immunisations whenever possible.
- Notification of measles and rubella should be done promptly on clinical diagnosis, not awaiting laboratory confirmation.

Further reading and information

Department of Health. *Immunisation against Infectious Disease 2006* (the 'Green Book'). Available at http://www.dh.gov.uk/en/Policyandguidance/Healthandsocialcaretopics/Greenbook/DH_4097254.

Department of Health latest evidence regarding MMR is available at www.mmrthefacts.nhs.uk.

http://www.hpa.org.uk/infections/topics_az/pregnancy/rashes/default.htm.

Case 9 A 68-year-old woman with fever and muscle pains

A 68-year-old woman attends your surgery complaining of 'the flu', fever and aching muscles.

What details would you like to elicit from the history?

History of presenting complaint

- Duration of fever and aches.
- Areas affected by aches.
- Anything that makes the aches better or worse.
- Any other family members and/or contacts affected.

Associated symptoms

- Appetite, nausea, vomiting or bowel upset?
- Sore throat or conjunctivitis?
- Current and recent general health.
- Cough, sputum, wheeze or shortness of breath?

Past medical history

- Immunisation history.

Drug history

- Prescribed or recent over-the-counter medication.
- Any allergies.

Social history

- Smoker, current or previous?
- Travel history.
- Occupation.

Further history reveals that she became unwell 3 days ago with a 'high' fever, general aches in her muscles and bones for which she has been taking paracetamol regularly with only partial relief. Her throat has been sore and she developed a cough yesterday, initially non-productive, but now productive of a little yellow sputum. She is feeling slightly short of breath and has attended because she thought that antibiotics, or perhaps the flu drug she has read about in the newspaper, might be of help.

She cares for her grandchildren, a 6-year-old, after school, and an 18-month-old, 3 days a week. Although the 6-year-old has been well, the 18-month-old was not able to attend nursery recently because of a cold. Because of her own illness, she has been unable to care for her grandchildren the last 3 days and knowing how her daughter and partner rely on her, she wants to get better as soon as possible. She was immunised against influenza in October (now January) and has previously had pneumococcal vaccine. She is a lifelong non-smoker, lives with her husband who is well and she has not been abroad for over 6 months.

On examination she is febrile (39 °C), looks unwell, is slightly short of breath at rest (respiratory rate 24 breaths/min) and coughing. Her throat is mildly inflamed, there is no cervical lymphadenopathy, and she has normal heart sounds, normal percussion and a clear chest.

What do you plan to do for this presumed influenza-like illness?

- *Discuss diagnosis.* This is likely to be a viral upper respiratory tract infection (URTI), due to influenza or one of many other respiratory viruses. With influenza, symptoms are often abrupt in onset with a systemic illness initially predominant (fever, headache, myalgia) and respiratory symptoms developing subsequently and predominating later (cough, sore throat, blocked or runny nose).
- *Discuss investigations.* These are not usually indicated for viral URTI.
- *Discuss management.* Advise continuing with regular paracetamol. Encourage maintenance of (non-alcoholic) fluid intake.

Infectious Disease: Clinical Cases Uncovered. By H. McKenzie, R. Laing, A. Mackenzie, P. Molyneaux and A. Bal. Published 2009 by Blackwell Publishing. ISBN 978-1-4051-6891-5.

Explain that you appreciate that she is not well and that with regular paracetamol her condition should improve within 2–3 days. Explain that should she deteriorate, she must make contact with the surgery or out-of-hours emergency care services.

What features should make her do this and why?

- Worsening fever or fever persisting more than the next 3 days.
- Worsening cough.
- Worsening breathlessness.
- Unable to maintain fluid input.
- Any evidence of confusion.

You tell her to explain this to her husband – for any of these symptoms, hospital admission might be necessary.

What about the 'flu drug' about which your patient is asking?

Zanamivir and oseltamivir can be used for treatment, within their licensed indications, during the influenza season. The Department of Health issues guidance each year as to when this has started.

Treatment is indicated for adults and children (>5 years, zanamivir; >1 year, oseltamivir) who can start treatment within 48 hours of the onset of symptoms and who have one of the following conditions that gives an increased risk of a more serious illness:

- Chronic respiratory disease.
- Significant cardiovascular disease.
- Chronic renal disease.
- Immunosuppression.
- Diabetes mellitus.

As your patient does not fall into any of these categories and has been ill for 3 days, there are two reasons why neither influenza drug is indicated.

What about antibiotics?

Antibiotics have no action against viruses and your patient has no good evidence for a bacterial infection requiring treatment.

What about microbiological investigations?

Influenza-like illnesses are very common in primary care. In most cases, symptomatic management with paracetamol alone, without any laboratory investigation, is appropriate. Investigation of all such patients would overwhelm laboratory diagnostic facilities and would seldom influence patient management, even if results were available within 1–2 days.

When might microbiological investigations be appropriate?

- A possible outbreak in a closed community.
- In immunocompromised patients.
- Unusual clinical presentation or serious systemic illness requiring hospital admission.

What diagnostic methods would be used when appropriate for viral respiratory infections?

For selected patients, establishing the cause of a presumed viral respiratory infection is appropriate. This is commonly done for hospitalised children by antigen detection with immunofluorescence (usually for influenza A, influenza B, adenovirus, respiratory syncytial virus, parainfluenza virus and metapneumovirus) and for immunocompromised patients by nucleic acid detection (for these organisms and also others, e.g. rhinovirus, coronavirus). These methods are preferable to culture, which is both slower and less sensitive, and to serology, which is also slower, less sensitive and available for fewer organisms (see Antigen detection and viral diagnosis in Laboratory diagnosis of infection, Part 1).

What are some complications of influenza?

- Bronchitis, especially in older patients or those with chronic obstructive pulmonary disease (COPD).
- Pneumonia, which can occur in the previously healthy but especially in those with pre-existing diseases. It may be due to the influenza virus itself (high mortality and death may be rapid, occurring even within a few hours of onset of the influenza illness) or bacteria (lower mortality than influenzal pneumonia and onset slower). However, *Staphyloccocus aureus* pneumonia is a well recognised complication of influenza and has a high mortality.
- Reye's syndrome, an encephalopathy with fatty liver degeneration. It has a high mortality in children with prior aspirin treatment. Aspirin is usually contraindicated for those under the age of 16 years.
- Diabetic ketoacidosis, even in those with mild diabetes.

What preventative measures are available for influenza?

Annual active immunisation

Vaccines containing inactivated components of influenza A and influenza B viruses are available. The composition of virus subtypes is reviewed each year by the World Health Organisation (WHO) to try to ensure a close match between the strains included in vaccines and the virus subtypes in current circulation.

How effective is influenza immunisation?

Protection is 70–80% when the included viral strains match well with the seasonal circulating viral strains.

For whom is influenza immunisation indicated?

Annual immunisation is recommended for all patients over 65 years of age. Immunisation is also recommended for patients with:

- Chronic respiratory disease, including asthma.
- Chronic heart disease.
- Chronic renal disease.
- Chronic liver disease.
- Diabetes mellitus requiring insulin or oral hypoglycaemic drugs.
- Immunosuppression.

Antiviral treatment of influenza

During the influenza season, oseltamivir can be used for prevention after exposure for selected at-risk individuals aged 1 year and over. It is not indicated for otherwise healthy people under 65 years of age.

What are the differences between an outbreak, an epidemic and a pandemic?

- *Outbreak:* cases of disease associated in time or location.
- *Epidemic:* a large but temporary increase in cases of a disease above that normally expected in a community or area.
- *Pandemic:* a worldwide epidemic (or at least affecting many countries).

Influenza can occur as sporadic cases, but seasonal influenza is an annual event that causes local outbreaks or more widespread epidemics. It is generally accepted that another pandemic of influenza is inevitable because of antigenic change in the influenza virus. There are three types of influenza virus: A, B and C. Antigenic change in influenza viruses can occur gradually (antigenic drift) or suddenly (antigenic shift). Only influenza A 'shifts', whereas A and B can 'drift'. Influenza B and C are human pathogens only, whereas aquatic birds are the reservoir of influenza A, and many birds and animals (including humans) can be infected. Pandemic influenza is due to influenza A.

What events will lead to an influenza pandemic?

- Antigenic evolution of an influenza virus to which humans have little or no immunity.
- Easy spread of this virus amongst susceptible humans.

Pandemics due to influenza have occurred periodically over the centuries, nine since 1700. In the 20th century there were three pandemics: in 1918/19, 1957 and 1968. Though it is impossible to know how soon the next will be, where it will start or how severe it will be, it is believed that it is a matter of 'when' and not 'if'. As it is over 40 years since the last pandemic, it is likely that current medical students and junior doctors will be working during the next one.

How will we cope with the next influenza pandemic?

This will be a major challenge for healthcare services. Many healthcare workers will be unwell and unable to work and others will have to care for ill dependents. Modelling suggests that within the UK, over the entire period of a pandemic, total healthcare contacts may increase from around 1 million during a 'normal' influenza season to around 30 million. As an example, if 25% of the population is infected, in a population of 100 000, this will result in:

	Total	During the peak week
Clinical cases	25 000	5500
GP consultations	7 130	1570
Hospital admissions	1 000	220
Deaths (if case fatality rate of 2.5%)	625	140

This would put major pressure on staffing and resources, e.g. maintaining infection control procedures, staffing intensive care, etc. It is essential that there is planning at national and local level to meet these demands.

Back to your patient

You decided that there was no indication for microbiological investigation. In view of her 18-month-old grandchild's illness, to what respiratory viruses is your patient likely to have been exposed?

• *Respiratory syncytial virus (RSV).* Nearly all children have been infected by the age of 2 years. RSV infection is seasonal with a winter peak every year in temporate climates. Illness ranges from mild symptoms to bronchiolitis or pneumonia, these being the commonest presentations of RSV that may require hospital admission. Re-infections occur throughout life with serious illness most likely in those with underlying lung disease or the immunocompromised.

• *Metapneumovirus (MpV).* Other than most being infected by 5 years, the symptomatology is clinically indistinguishable from RSV.

• *Parainfluenzaviruses (PIVs).* After RSV, these are the commonest cause for hospitalisation of young children for respiratory tract infections, presenting mainly as croup, bronchiolitis or pneumonia. As with RSV and MpV, there is seasonality (e.g. PIV type 1 in winter, PIV type 3 in summer), and re-infections occur which in adults usually cause URTI.

• *Rhinovirus.* With over 100 serotypes and re-infection with the same serotype possible, these are the major causes of the common cold with increased morbidity in the elderly.

What is the reason for enquiring about foreign travel?

Malaria and some other tropical infections can present with a flu-like illness and without the 'pointer' of the travel history are unlikely to be considered within the differential diagnosis.

Progress and outcome

Your patient makes a gradual recovery over the following week from the flu symptoms, though she still feel tired and it is a further 3 weeks before she feels fully better.

CASE REVIEW

A 68-year-old woman presented with a community-acquired 'flu-like' illness. Her 18-month-old grandson recently suffered from cold which is a common symptom for a number of different viral URTIs such as influenza, RSV, human metapneumovirus, PIV and rhinovirus. The grandmother could have easily contracted one of these viruses due to close contact. Laboratory investigations were not considered necessary and she made a slow and complete recovery with symptomatic treatment only.

KEY POINTS

- Many viral illnesses of the respiratory tract can lead to severe symptoms, particularly in patients who have other predisposing factors.
- Influenza immunisation is the most important means of preventing influenza infection.
- Guidelines for influenza prevention and treatment are available from local and national health bodies. These guidelines get regularly updated.
- Preparations for the next influenza pandemic are important.

Further reading and information

Antivirals in the prevention and treatment of influenza: http://www.dh.gov.uk and http://bnf.org/bnf/bnf/current/104945.htm.

Infection control precautions for seasonal, avian and pandemic influenza: http://www.hps.scot.nhs.uk/resp/rpicsi.aspx?subjectid=AA.

Influenza vaccination: http://www.dh.gov.uk/en/Policyandguidance/Healthandsocialcaretopics/Greenbook/DH_4097254.

Pandemic influenza: http://www.dh.gov.uk/en/Publicationsandstatistics/Publications/PublicationsPolicyAndGuidance/DH_073185 (Department of Health, England); http://www.hpa.org.uk/publications/2006/pandemic_flu/pandemic_flu_plan.pdf (Health Protection Agency, England); http://www.brit-thoracic.org.uk (British Thoracic Society).

Parainfluenzavirus: http://www.hps.scot.nhs.uk/resp/parainfluenzavirus.aspx?subjectid=182.

Respiratory syncytial virus: http://www.hps.scot.nhs.uk.

Case 10 A 35-year-old teacher with fever and chills on return from Malawi

A 35-year-old teacher was admitted to hospital with a 2-day history of fever and chills 7 days after returning from Malawi. She had been there for 1 month participating in an educational exchange programme.

Suggest some possible causes for this woman's symptoms

1 Viral infection such as influenza.
2 Urinary tract infection.
3 Pneumonia.
4 Malaria.
5 Typhoid fever.

What details would you like to elicit from the history?

Presenting complaint and history of presenting complaint

• *Duration.* A short history is suggestive of an acute infection, whereas a longer history of weeks and months is more suggestive of chronic infection or connective tissue disease. The minimum time from a mosquito bite to the clinical development of malaria is 7 days. Patients often present with falciparum malaria 1–4 weeks after exposure, although for non-falciparum malaria this may be many weeks or possibly months later.

• *Travel history.* As people travel to more exotic places, it is important to be confident of where the patient has been, e.g. Equatorial Guinea (Central Africa) and Papua New Guinea (Far East) are different places. There may have been a particular outbreak in one specific area recently.

• *Activities.* Many tropical countries have endemic disease to which the traveller may or may not have been exposed depending on their activities while there (e.g. schistosomiasis after swimming in fresh water).

• *Contact with sick people.* It is important to ask whether she has had contact with others who have had similar illness (e.g. respiratory symptoms, jaundice, fever), which may suggest a possible outbreak.

• *Chemoprophylaxis.* Enquiry should be made regarding chemoprophylaxis against malaria or other conditions while travelling. In particular, the name of the drug and some assessment of adherence with the regimen are important. While no anti-malarial regimen is 100% effective in preventing malaria, compliance with a recommended regimen would make malaria less likely.

• *Pre-travel immunisation.* Immunisation against typhoid, polio, hepatitis A, hepatitis B, diphtheria, tuberculosis, Japanese B encephalitis, yellow fever and rabies make infection with these conditions much less likely.

• *Other symptoms.* All the above questions are helpful to 'set in the background.' It is now important to determine whether she has other specific or 'localising symptoms' which might point to a particular diagnosis. Remember that about 50% of travellers return with a cosmopolitan infection (e.g. pneumonia or influenza) which may have been contracted in their home town or on the aircraft rather than in a developing country. A cough productive of green sputum, pleuritic chest pain and dizziness may suggest pneumonia. Malaria and typhoid may also present with a cough, while headache, photophobia and neck stiffness suggest meningitis. Headache is a common feature of many acute infections. Ear, nose and throat symptoms generally point to an upper respiratory tract infection. Diarrhoea, often watery, with crampy abdominal pain would suggest gastroenteritis, but loose stool is common with many acute infections. Constipation is a feature of typhoid fever. Urinary frequency, dysuria and offensive urine point to a urinary tract infection, although dark, offensive, concentrated urine can occur with any

Infectious Disease: Clinical Cases Uncovered. By H. McKenzie, R. Laing, A. Mackenzie, P. Molyneaux and A. Bal. Published 2009 by Blackwell Publishing. ISBN 978-1-4051-6891-5.

febrile condition. Especially dark 'black water' urine may be a feature of severe malaria.

• *Pregnancy.* It is also important to enquire about the possibility of pregnancy in a woman of childbearing age, especially in the returning traveller. This is because malaria is likely to be more severe in pregnant women and the choice of treatment is more restricted.

Past medical history

• Has the patient ever had a similar illness (e.g. malaria), and if so how and where was it diagnosed and treated? A patient with immunodeficiency (e.g. HIV infection, leukaemia, lymphoma, myeloma) is at risk of particular opportunistic infections which should be considered.

• Chronic medical disorders such as diabetes mellitus, chronic liver and kidney disease are associated with a significant infection risk, as are patients on immunosuppressive therapy (e.g. steroids, cyclosporin, cyclophosphamide).

• It is important to take a full past medical history as all sorts of medical conditions are relevant when planning therapy, e.g. atrial fibrillation, epilepsy.

Drug history

It is important to be aware of current or recent antibiotic or anti-malarial medication as this may lead to false-negative laboratory results. Treatment failure raises the possibility of resistance.

Social history

The patient's occupation is often relevant and should always be sought. On this occasion, exposure to children, not only in the UK, but also in Malawi, may point to a respiratory viral infection. Heavy alcohol or drug misuse may increase the risks of certain infections such as HIV. Household contacts are important as they may also be developing or have had a similar infection.

The history is of acute onset fever, sweats, malaise and fatigue with no localising symptoms. She knows that she is now 8 weeks pregnant, having had a positive test the week before leaving for her 1 month trip to Malawi. She did not take anti-malarial prophylaxis because of conflicting advice about the need for prophylaxis and its safety in pregnancy. She received all the appropriate recommended immunisations before travel. She has no other past medical history and her husband is well.

Time to review your differential diagnosis

1 *Viral infection.* An upper respiratory tract infection may be characterised by a runny nose/blocked nose and sore throat. In contrast, influenza usually presents with a more severe malaise, fatigue, fever and aches and pains, and less predominant respiratory symptoms initially (see Case 9).

2 *Urinary tract infection.* The absence of any urinary symptoms, e.g. dysuria, frequency, loin pain, does not rule out this diagnosis, but makes it less likely.

3 *Pneumonia.* While cough, green sputum, dyspnoea or pleurisy may occur, many patients simply present with fever and malaise.

4 *Malaria.* This would seem to be the most likely diagnosis and also the most important to consider in a traveller returning from southern Africa. Malaria presents with a non-specific 'flu-like' illness, possibly with a dry cough, dark urine and aching joints.

5 *Typhoid fever.* This also presents with a non-specific febrile illness and can be difficult to diagnose clinically. It is less likely as she has had a typhoid vaccination.

What would you look for on examination?

As the initial temperature measurement may be normal, several subsequent measurements are necessary to document the fever or, if present, hypothermia. The presence of hypotension should be sought as this may represent hypovolaemia/dehydration, or even septic shock. Tachycardia is common with many febrile infections, although a 'relative bradycardia' may be seen with typhoid fever. Hepatosplenomegaly is occasionally seen in acute malaria, but is more common in indigenous people in endemic areas. Signs of consolidation may be present in pneumonia, but widespread crackles may be present in severe malaria complicated by pulmonary oedema/adult respiratory distress syndrome. Drowsiness points to severe, life-threatening infection (e.g. cerebral malaria, meningitis, septicaemia). Full examination including the cardiovascular system, central nervous system and locomotor system is indicated in order not to miss subtle signs that may point to an alternative diagnosis.

The patient was alert and orientated with a pulse of 100 beats/min, BP 120/80 mmHg and temperature 38.8 °C. Chest, cardiovascular system, abdomen and central nervous system were normal.

What is the likely diagnosis?

Malaria, typhoid or a viral illness.

What would you do now?

As the patient is not critically ill, it is possible to perform the necessary baseline investigations first to try to confirm the diagnosis before commencing treatment.

What tests would you do?

Malaria blood film

A thick blood film is used to look for any malaria parasites that may be present and the thin film used to identify the species and estimate the parasitaemia (percentage of red cells containing one or more parasites). An example of a thin film is shown in Plate 10. The laboratory may use a malaria antigen detection test or a quantitative buffy coat fluorescence test on the blood as initial screening tests.

Full blood count

The platelet count is often helpful in assessing fever in the returning traveller as thrombocytopenia (<100 × 10^9/L) is generally present in patients with malaria. Thrombocytopenia is also present in many viral and some severe bacterial infections, so it is not specific to any one infection. One would expect the haemoglobin to be normal or slightly low in the conditions being considered in the differential diagnosis for this patient. A particularly low haemoglobin (<8 g/dL) is likely to be due to severe haemolysis and may require transfusion support. The white cell count is usually normal in malaria and it may be low in typhoid fever. There may be a neutrophil leucocytosis with pneumonia and lymphopenia with many viral infections.

Urea and creatinine levels

Elevated urea and creatinine is suggestive of renal dysfunction and may occur in many conditions. However, in malaria, this would be an adverse prognostic sign. Bilirubin and lactic dehydrogenase (LDH) are likely to be elevated in conditions characterised by haemolysis (e.g. malaria) and the liver enzymes may be abnormal in many conditions.

Blood culture

In many bacterial infections, particularly typhoid fever and pneumonia, this will be the only way of confirming the diagnosis and often multiple sets of blood cultures are sent before commencing therapy.

Urine culture

This might confirm a urinary tract infection, although it may also be possible to recover the organism causing a bacteraemia (e.g. *Salmonella typhi* in typhoid fever).

Chest X-ray

A chest X-ray may be abnormal in pneumonia or reveal a complication in other conditions.

Time to review the patient

Two hours later the following results are back from the laboratory:

		(Normal range)
Haemoglobin	*12.2 g/dL*	*(11.5–16.5)*
White blood cell count	*9.4 × 10^9/L*	*(4–11)*
Platelet count	*58 × 10^9/L*	*(150–400)*
Blood film	*Ring forms of Plasmodium falciparum*	
Parasitaemia	*2%*	
Serum urea	*7.0 mmol/L*	*(2.5–7.5)*
Serum creatinine	*95 μmol/L*	*(60–110)*
Serum total bilirubin	*30 μmol/L*	*(1–22)*
Serum alanine aminotransferase	*49 U/L*	*(5–35)*
Serum gamma glutamyl transferase	*65 U/L*	*(4–35)*
Serum alkaline phosphatase	*115 U/L*	*(45–105)*
Serum C-reactive protein	*95 mg/L*	*(<10)*
Pregnancy test	*Positive*	
Chest X-ray	*Normal*	

What is your diagnosis now?

Plasmodium falciparum malaria in a 35-year-old female who is 8 weeks' pregnant.

Each year about 1700 cases of malaria occur in the UK, 1300 in the USA and 3000 in France. In the UK about 75% of these cases are caused by *P. falciparum* which produces the most severe form of malaria. Five to 16 deaths occur annually in the UK and are nearly always in cases of falciparum malaria.

Four main species of malaria parasites infect humans after a bite of an *Anopheles* mosquito: *P. falciparum*, *P. vivax*, *P. ovale* and *P. malariae*. *P. falciparum* and *P. vivax* are the most common. Falciparum malaria predominates in sub-Saharan Africa and vivax malaria in the Indian subcontinent, Mexico, Central America and China. Both species occur in South East Asia and South America.

The illness due to non-falciparum species tends to be less severe. Note that infection with *P. vivax* or *P. ovale* may relapse several months after apparently successful treatment, as both species can persist as hypnozoites in the liver.

How should this patient be treated?

Because of concerns about treatment failure due to low-grade quinine resistance, the British guidelines (2007) for the treatment of malaria, suggest the use of two drugs for the treatment of falciparum malaria, e.g. quinine and doxycycline (or clindamycin) or Malarone (atovaquone and lumefantrine) or Riamet (artesunate and primoquine).

All these would be suitable for uncomplicated malaria. However, complicated disease, which includes parasitaemia of >1% and pregnancy, would require treatment with intravenous quinine plus either doxycycline or clindamycin. Since doxycycline is contraindicated in pregnancy (congenital abnormalities in the unborn child), this patient should be treated with intravenous quinine (with electrocardiogram (ECG) and glucose monitoring) and intravenous clindamycin. She should also receive intravenous fluid and daily monitoring of her parasite count, blood count and biochemistry.

Outcome

Following treatment with IV quinine and clindamycin, her fever settled within a day and her parasites cleared within 3 days. She was then changed to oral quinine and clindamycin and was discharged after completing 7 days of treatment. She made a full and uncomplicated recovery.

Should she have taken malaria prophylaxis for her trip to Malawi?

The best approach to prevent malaria in pregnancy would have been cancelling the trip to Africa and this option should always be considered. If, however, the trip is essential, anti-malarial prophylaxis would always be indicated for pregnant women. The two suggested prophylatic regimens, Malarone and mefloquine, are not known to be harmful in pregnancy. The risks of malaria to the mother and the unborn child are, however, significant. Doxycycline should not be used in pregnancy.

CASE REVIEW

A 35-year-old woman has returned from Malawi with a febrile illness. She had not taken malaria prophylaxis because of concerns about its safety in pregnancy – probably misplaced concerns given the balance of risk. She is 8 weeks' pregnant and blood tests confirmed the diagnosis of malaria. She was treated with a combination of quinine and clindamycin and made a successful recovery.

KEY POINTS

- It is important to take an accurate travel history from returning travellers presenting with a febrile illness.
- Malaria (blood film) and typhoid (blood cultures) are important differential diagnoses to exclude in patients who have returned from a malarious area.
- There are four different forms of malaria (*Plasmodium* species), all spread by the *Anopheles* mosquito. *P. falciparum* causes the most severe form of the disease.
- Guidance on immunisations and malaria prophylaxis should be checked prior to travel (see websites below).
- Certain antibiotics are contraindicated in pregnancy. Others are used only when benefits outweigh the risks.

Further reading and information

Centre for Communicable Diseases in the USA has an extensive range of advice for travellers available at www.cdc.gov/travel.

UK malaria treatment guidelines prepared by the British Infection Society are available at http://www.britishinfectionsociety.org.

UK travellers' advice is available at www.fitfortravel.nhs.uk.

Case 11 A 32-year-old man with night sweats and fatigue

A 32-year-old man was referred to hospital complaining of recurrent fevers. These had been present for over a month and associated with drenching sweats at night that had been so bad that he had needed to change his night clothes on several occasions. In addition he described fatigue, weight loss and loss of appetite.

How should you approach this problem?

Perhaps the first thing to do is to establish that the fevers actually do exist. This can be achieved by providing the patient with a thermometer to monitor his own temperature or, preferably, by admitting to hospital for a period of observation and monitoring.

The presence of significant fevers (>38 °C) was thus documented and the next step was to get some more information from the patient. One of the most important helpful things to know at this stage is the *duration* of fevers. This patient gave a history that dated back more than 4 weeks. A patient with a prolonged unexplained fever is described as having a fever (or pyrexia) of unknown origin (FUO). Strictly speaking this term should be applied to patients who have had more than 3 weeks of fevers with temperatures of >38.3 °C recorded on several occasions *and* with no diagnosis reached after 1 week of hospital assessment. Most febrile illnesses last less than 3 weeks or are diagnosed within a few days of admission to hospital, so patients with a FUO as defined above represent a particular challenge.

What next?

Take a very thorough history. You are unlikely to get to the root of a problem like this with a 10-minute consultation – the clues to the diagnosis may be subtle and easily overlooked. Do not feel embarrassed about going back and re-taking the history in case you have missed anything. Indeed, many doctors will ask for a second opinion on such patients to get a fresh perspective on what can be a major diagnostic challenge. Some of the important aspects of an FUO history are summarised in Table 11.1.

The patient worked as a postman. He described intermittent joint discomfort in the knees, ankles and wrists. He had been aware of occasional rashes which he said were pink and flat but these tended to come and go, usually lasting no more than a couple of days. His appetite had been poor and his weight had fallen by 7 kg in the last month. There was no history of travel abroad and his only medication was paracetamol, which he took for the joint pains. He had no history of drug misuse. He had received a course of amoxicillin for 1 week when he first went to his family doctor but the fevers had not responded to this. He had a past history of appendicectomy aged 10 years and there was no family history of joint disease or febrile illnesses.

What about the examination?

Just as with the history, the examination should be unhurried and thorough in these patients. The findings may be subtle and sometimes changing, e.g. a new murmur in the patient with endocarditis. Examination of the skin, lymph nodes, oral cavity and fundi should not be overlooked. Some possible examination findings and their significance are summarised in Box 11.1.

Examination in this case revealed bilateral small knee effusions and moderate splenomegaly. There was no rash and no lymphadenopathy.

Where do you go from here?

Once you realise that your patient has an FUO you can start to think about diagnostic categories. Broadly speaking, there are four of these:

Infectious Disease: Clinical Cases Uncovered. By H. McKenzie, R. Laing, A. Mackenzie, P. Molyneaux and A. Bal. Published 2009 by Blackwell Publishing. ISBN 978-1-4051-6891-5.

Table 11.1 Important features of a history of a fever of unknown origin.

History	Examples
Pattern of fever	Fevers may occur every 2 or 3 days with some forms of malaria
Other symptoms	Joint symptoms, mouth ulcers, rashes, diarrhoea, weight loss, cough, headaches
Occupation	Febrile reactions can occur to some chemicals or metal fumes
Travel	May be relevant to increased risk of malaria, tuberculosis or HIV
Past history	Sometimes the FUO can be relapsing, e.g. adult Still's disease
Family history	Familial Mediterranean fever
Social history	Hobbies, pets, illicit drug use especially drug injection
Drug therapy	Many prescribed drugs can produce idiosyncratic febrile reactions

Box 11.1 Importance of different examination findings in this case

Examination finding	Possible relevance
Heart murmur	Endocarditis
Joint swelling, tenderness	Connective tissue disease, e.g. systemic lupus erythematosus (SLE)
Rash	Still's disease, vasculitis, endocarditis
Fundal haemorrhages/exudates	Endocarditis, cytomegalovirus retinitis
Lymphadenopathy	HIV disease, lymphoma
Splenomegaly	Lymphoma, adult Still's disease, endocarditis
Tender temporal artery	Temporal arteritis (giant cell arteritis)
Oral ulceration	Crohn's disease

- Infection.
- Malignancy.
- Multisystem disease (includes rheumatologic diseases/vasculitis and granulomatous diseases, e.g. sarcoidosis).
- Other – including undiagnosed and fabricated fevers.

As diagnostic techniques have improved over the years we have become better at diagnosing many infections and malignancy. Recent studies have suggested that about a quarter of patients will have an infection, around a third each will have a multisystem disorder or other diagnosis, and about one-tenth will have malignancy as the cause of their FUO. The answer in this case will not come from the history and examination alone so further investigations are needed.

Investigations of an FUO

By the time they have been referred to hospital many patients will already have had some tests performed. All patients should have the following:

- A full blood count.
- Biochemistry (urea and electrolytes and liver function tests).
- Erythrocyte sedimentation rate (ESR) and C-reactive protein.
- Urinalysis for blood and protein.
- Chest X-ray.

Other investigations ought to be guided by the history and examination findings.

The results for this patient were:

		(Normal range)
Haemoglobin	*11.1 g/dL*	*(12–16)*
White blood cell count	*18.0 × 10^9/L*	*(4–11)*
Serum alanine aminotransferase	*110 U/L*	*(5–35)*
Serum C-reactive protein	*155 mg/L*	*(<10)*
ESR	*80 mm/h*	
Chest X-ray	*Normal*	
Urinalysis	*Normal*	

The raised white cell count, ESR and CRP are all in keeping with inflammation but give no indication of the cause. The low haemoglobin most probably reflects the chronic nature of the inflammation – a well recognised cause of anaemia.

So what is likely to be going on here and what other tests may be of help?

Perhaps it would help to think of the problem in the broad diagnostic groups listed earlier.

- *Infection.* We have not excluded infection so we need to start by getting some blood cultures and perhaps consider an echocardiogram to look for evidence of endocarditis (the enlarged spleen might go with this).

• *Malignancy.* Splenomegaly can be a feature of lymphoma. Further imaging is needed to exclude a malignancy here – a CT scan of the chest and abdomen would be a good way to look for lymphadenopathy or tumour disease. Check the serum LDH as this is often elevated in lymphoma.
• *Multisystem disease.* The joint symptoms raise the possibility of a rheumatological disorder and an enlarged spleen can occur in some such conditions such as adult Still's disease. We need to test the patient's blood for rheumatoid factor and autoantibodies such as antinuclear factor.
• *Other.* The patient's stay in hospital has allowed objective measurement of his temperature so there is no doubt that it is real. There is nothing in his drug or occupational history of note and no family history to suggest a familial fever. The group of 'other' diagnoses are diverse and include such things as Crohn's disease and multiple pulmonary emboli but there are no localising symptoms in this case to suggest such possibilities.

Five consecutive blood cultures taken at times of fever were negative. The patient's CT scan of the chest and abdomen revealed splenomegaly but no significant lymphadenopathy. The rheumatoid factor and autoantibodies were negative and the LDH was normal. No vegetations were seen on the echocardiogram.

What should we do for this man?

A trial of antibiotics had been given by the family doctor with no benefit and there is no evidence of infection. As far as possible we have excluded malignancy. The joint pains, enlarged spleen, intermittent rash and raised white cell count would all support a diagnosis of Still's disease. Furthermore, this condition is not associated with positive rheumatoid factor or autoantibodies.

Given these suspicions it was decided to start the patient on a therapeutic trial of steroids – an appropriate treatment for Still's disease. He received 40 mg of prednisolone daily and within a week his fevers and joint pains had resolved, his liver function had returned to normal and his anaemia improved. The patient was discharged home with out-patient follow-up to be undertaken by the rheumatologists.

Adult Still's disease

This is an inflammatory disorder of unknown cause. Fevers are usually a very prominent feature so it is no surprise that these patients are often thought to have an infection initially. As well as fevers, patients tend to have arthralgia or arthritis and a raised white cell count. When febrile, patients often report a pink rash. Other features may include an enlarged liver and/or spleen and lymphadenopathy. The rheumatoid factor and antinuclear factor are negative. Sometimes the patient will recover spontaneously from an episode but in other cases the symptoms can come and go or become chronic. A number of drugs have been used to treat this condition including non-steroidal anti-inflammatory drugs, steroids and methotrexate. The choice of drug depends on the severity and features of the disease.

CASE REVIEW

A 32-year-old man was admitted to hospital for investigation of prolonged fever. Attempts were made to find out the cause of this, traditionally classified as fever or pyrexia of unknown origin if a diagnosis has not been made within 1 week of admission. A full and detailed history was taken and a thorough examination was carried out. Infection and malignancy were excluded and based on the symptoms of fever with joint pains and a skin rash, a provisional diagnosis of Still's disease was made. The patient benefited from a therapeutic trial of steroids and was referred to a rheumatologist for further treatment.

KEY POINTS

- The patient should have significant undiagnosed fevers for over 3 weeks before the term FUO is applied.
- Infection is the cause of an FUO in around 25% of cases.
- A detailed history and examination are key to pointing you in the right direction of the diagnosis.
- Sometimes you need a trial of therapy to get the diagnosis since not all diseases have diagnostic tests.

Further reading and information

There are many reviews on fevers of unknown origin. One example that may be of interest to readers is available online at http://www.aafp.org/afp/20031201/2223.html.

Case 12 A 41-year-old male with a red, hot and swollen left lower leg

A 41-year-old male engineer was sent to hospital by his general practitioner with a 1-day history of an acute, painful, red, hot, swollen left lower leg. He also complained of general malaise, fatigue and fever.

What are the possible causes of this presentation?

1 Cellulitis.
2 Erythema nodosum.
3 Erythema migrans (Lyme disease).
4 Trauma.
5 Deep venous thrombosis.

What details would you like to elicit from the history?

Presenting complaint and history of presenting complaint

- Any previous episodes – both erythema nodosum and cellulitis can be recurrent.
- Any other sites affected – cellulitis would usually be limited to one area, whereas erythema nodosum may occur on both shins and possibly the arm.
- Trauma or injury – was there a precipitant either recently or in the past?
- Problems with the feet (e.g. athlete's foot, fungal nail infection – which can provide an entry point for infection).
- Any history of tick bite or potential for tick exposure that might suggest Lyme disease.

Other symptoms

- Fever.
- Rigors.
- Nausea, vomiting and headache.
- Vomiting.
- Headache.
- Fatigue.
- Malaise.

Past medical history

- Cellulitis may be recurrent and therefore a previous history of a similar episode is a vital piece of information.
- Past history of leg swelling, e.g. lymphoedema, would predispose to cellulitis.
- Has there been trauma with or without puncture of the skin and subsequent introduction of infection? Blunt trauma is also a risk for cellulitis.
- Previous surgery (e.g. harvesting of leg veins for coronary artery bypass surgery) is a risk for subsequent cellulitis.
- Previous bony injury (e.g. fracture) – this is a risk for subsequent osteomyelitis, which may present with a superficial cellulitis.
- Diabetes mellitus – there is an increased risk of cellulitis because of hyperglycaemia, peripheral neuropathy and peripheral vascular disease.

Drug history and allergies

- If this is a recurrence, a previous response to a particular antibiotic regimen may guide treatment on this occasion. Conversely, treatment failure with a previous regimen would be important to establish.
- A history of antibiotic allergies should be actively sought. About 10% of patients claim to be allergic to penicillin, but only 9% have a true allergy. Symptoms of nausea, vomiting or diarrhoea can be due to the drug but can also be due to the infection that is being treated. Patients with a true allergy to penicillin should not receive any penicillin. Cephalosporins are chemically related to penicillin, but only 5–10% of those allergic to

Infectious Disease: Clinical Cases Uncovered. By H. McKenzie, R. Laing, A. Mackenzie, P. Molyneaux and A. Bal. Published 2009 by Blackwell Publishing. ISBN 978-1-4051-6891-5.

penicillin are allergic to cephalosporin. Therefore, unless the patient has a true immediate hypersensitivity reaction (e.g. anaphylaxis, bronchospasm, mucosal swelling or immediate rash), cephalosporin may be a treatment option.

- A history of intolerance to other antibiotics should also be sought.

Family and social history

- Occupation – some occupations require particular footwear (e.g. boots, training shoes) which may predispose to athlete's foot and therefore to cellulitis. A history of hill-walking or working with deer may point to tick bites and Lyme disease.
- A history of injection drug use (e.g. heroin) represents an important predisposing factor for skin and soft tissue infection and should not be missed.

The patient recalls having one similar episode of this 3 months ago which resolved spontaneously after about 4 days. Further history revealed that the patient plays amateur league football two or three times per week. He does not recall any blunt injury to the left shin before the onset of these symptoms. His only past history is of a fractured left lateral malleolus at aged 19 years which did not require orthopaedic surgery. There is no family history of note. He is on no regular medication and does not take drugs.

Time to review your differential diagnoses

1 *Cellulitis.* This would seem to be the most likely diagnosis given the location on the lower limb. He may have athlete's foot, a common finding in footballers, as a predisposing factor. Injection drug use-related skin infection often occurs in the groin due to intravenous injection into the femoral vein, but may occur in the thighs and shins due to 'skin popping'.

2 *Erythema nodosum.* This condition is characterised by multiple, slightly raised pink/purple warm, tender round lesions, 1–4 cm in diameter, that may be mistaken for cellulitis. However, the multiple, discrete nature of these lesions, which may also be found on the arms, usually differentiates this condition from cellulitis.

3 *Lyme disease.* Erythema migrans is a progressively enlarging pink/red, warm, non-tender, non-raised macule, 1–30 cm in size, at the site of a previous tick bite and represents acute Lyme disease due to infection with *Borrelia burgdorferi.* The initial lesion is often on the lower leg as this is the most likely site of a tick bite, but it may occur anywhere on the body. The lesions may be multiple as the infection becomes bacteraemic.

4 *Trauma.* Can appear as a slightly pink, warm, raised swelling and there would usually be a history of trauma at the site.

5 *Deep venous thrombosis.* The whole of the leg would appear swollen and warm, but one would expect little colour change.

What are the key things to look for on examination?

- The site, size, colour, tenderness and temperature of the affected area.
- Whether the lesion is single or multiple.
- Examination of the feet for evidence of athlete's foot or fungal nail infection.
- Inspection of the legs and arms for evidence of injection drug use.
- An assessment of arterial circulation in the leg and any evidence of leg swelling to suggest deep venous thrombosis or lymphoedema.
- Any evidence of peripheral neuropathy.

On examination there was an extensive area of swelling, erythema and tenderness over the left leg and foot. His temperature was 37.8 °C and BP 135/80 mmHg. The skin, neurology and pulses in the lower limbs were normal. There were no other skin lesions. His left lower leg was slightly swollen compared to the right.

What is the likely diagnosis?

Cellulitis (Plate 11). is the most likely diagnosis, although he does not have any predisposing factors other than a previous left lateral malleolus fracture, which could serve as a nidus for a chronic osteomyelitis.

What are the common bacterial causes of cellulitis?

- *Streptococcus pyogenes.*
- *Staphylococcus aureus.*
- Group C and group G streptococci.

What are some less common bacterial causes of cellulitis?

- *Pasteurella multocida* – you would not normally think of this unless there was a history of animal contact. This might result from a bite or a pet licking a wound!

- *Aeromonas hydrophila or Vibrio vulnificus* – these infections are usually associated with exposure to water following skin trauma, e.g. an injury from a fish hook.

What investigations would you do?

Box 12.1 lists some helpful investigations.

Results

- *The full blood count, urea and electrolytes, and liver function tests were normal. The CRP was 35 mg/L and the ASO titre was not elevated.*
- *The X-ray of the left ankle was reported as 'Old fracture of lateral malleolus with no features of underlying osteomyelitis'.*
- *The Doppler ultrasound scan was normal with no evidence of deep venous thrombosis.*
- *Blood cultures were reported negative after 7 days' incubation.*

How would you treat this patient?

Milder forms of cellulitis can be treated with oral antibiotics in the community. The extent of this area of cellulitis suggests that the patient will require intravenous antibiotic therapy. An initial intravenous regimen such as benzylpenicillin (to treat *Streptococcus pyogenes*) and flucloxacillin (to treat *Staphylococcus aureus*) is commonly used. Intravenous therapy is often continued until the majority of the erythema and tenderness have dissipated, often after 4–7 days.

Commonly used alternative treatments for those with a penicillin allergy include ceftriaxone, teicoplanin and vancomycin. Clindamycin is a good alternative for cellulitis since it penetrates well into the tissues and covers most strains of *Staphylococcus aureus* and *Streptococcus pyogenes*. However, it is used with caution because of its association with *Clostridium difficile*-associated diarrhoea. New second-line Gram-positive antibiotics such as linezolid and daptomycin are also available for use in the case of penicillin allergy or treatment failure with a penicillin. Patients are often 'switched' to 'follow-on' oral therapy after the intravenous phase. Suitable oral drugs include penicillin V, amoxicillin, flucloxacillin, co-amoxiclav, clindamycin and linezolid.

Box 12.1 Useful investigations in this case

Possible investigations	Comments
Full blood count	May have an elevated white cell count
Urea and electrolytes and liver function tests	Expected to be normal unless the patient is very unwell
C-reactive protein (CRP)	Likely to be elevated if there is an infection
Blood cultures	Helpful if positive, but seldom are in cellulitis
Anti-streptolysin O (ASO) titre	May be positive in the convalescent phase of an acute streptococcal infection, but tends not to be high in cellulitis
X-ray of the left ankle	Look for evidence of chronic osteomyelitis in the left lateral malleolus
Doppler ultrasound scan	Would rule out a deep venous thrombosis

Progress

The patient responded well to a 1-week course of intravenous benzylpenicillin and flucloxacillin followed by a further week of oral penicillin V and flucloxacillin. He remained well for 3 months, but then re-presented with a further similar episode. This time he was treated with intravenous ceftriaxone used on a self-administered out-patient basis for 2 weeks, with complete resolution of his infection. At that stage a bone scan ruled out the possibility of osteomyelitis in the left ankle as an underlying cause. Unfortunately he suffered yet another episode several months later which once again responded to ceftriaxone.

He has enquired about what could be underlying this and what could be done to prevent further episodes. What do you advise?

Your patient has now required treatment on three occasions for cellulitis within a year. Consideration should be given to a trial of long-term oral antibiotic prophylaxis. There are no studies on which to base a recommendation, but options include penicillin V, amoxicillin, clindamycin, clarithromycin and co-trimoxazole. As his left leg is slightly swollen, it may be that he has secondary lymphoedema, which represents a risk of recurrent cellulitis. The use of a compression stocking on the affected leg is often recommended and can reduce the frequency of episodes of cellulitis.

Outcome

The combination of long-term oral penicillin V and compression stocking subsequently kept him free of any further episodes of cellulitis.

Complicated skin infections

Before leaving this case it is worth mentioning two presentations of skin infection that differ from typical cellullitis.

• Some strains of *Staphylococcus aureus* (usually community-acquired MRSA strains) are capable of producing a toxin known as Panton–Valentine leukocidin (PVL). The toxin is associated with an increased risk of skin infection, including recurrent skin abscesses, boils and necrotic infections.

• Necrotising infections (cellulitis or fasciitis) are associated with extensive tissue necrosis and a high mortality. These conditions can be caused by a variety of single organisms or a mixed infection, but *Streptococcus pyogenes* is one well recognised cause. In patients with necrotising fasciitis there may be relatively little superficial evidence to suggest cellulitis but the rapidly spreading infection in the deeper soft tissues and fascia leads to the patient becoming extremely unwell in a matter of hours. Aggressive surgery with removal of necrotic tissue needs to be combined with antibiotic and supportive treatment to offer any chance of survival.

CASE REVIEW

A 41-year-old man was admitted with a painful left leg. A diagnosis of cellulitis was made and he was treated with intravenous benzylpenicillin and flucloxacillin and later switched to oral antibiotics. However, his cellulitis recurred twice within a year and he was prescribed long-term oral penicillin V as a prophylactic agent.

KEY POINTS

• There are several causes of acute pain in the extremities associated with redness and swelling.

• Antibiotics used in the treatment of cellulitis should target the two most common bacterial causes – *Streptococcus pyogenes* and *Staphylococcus aureus.*

Further reading

Swartz MN. Clinical practice. *Cellulitis N Engl J Med* 2004; **350**(9): 904–12.

Stevens DL *et al.* Practice guidelines for the diagnosis and management of skin and soft tissue infections. *Clin Infect Dis* 2005; **41**: 1373–406.

Case 13 An 18-year-old female student with vaginal discharge

An 18-year-old female student attends your surgery complaining of a white, frothy vaginal discharge with a severe itch. One week ago she was seen by another GP in your practice complaining of urinary frequency and dysuria. A urinary tract infection had been suspected on clinical grounds and she was treated with co-amoxiclav for 7 days. These symptoms resolved and she says the current problem 'feels different' from the previous presentation.

What are the common causes of vaginal discharge?

Table 13.1 lists some of the common causes of vaginal discharge.

1 *Bacterial vaginosis (BV).* This is the commonest cause of vaginal discharge. It is a syndrome that is defined by mainly clinical criteria and diagnosis is based on the presence of any three of the following:

- The presence of a thin, homogenous vaginal discharge.
- The production of a 'fishy smell' on addition of potassium hydroxide to the discharge.
- Vaginal discharge pH > 4.5.
- The presence of 'clue cells' on microscopy.

The organism most commonly associated with BV is *Gardnerella vaginalis*, but it is not diagnostic of the condition and the aetiology has not been fully established. In BV there is a switch from a predominant flora of lactobacilli to an anaerobic flora with 'clue cells' seen on Gram stain. Clue cells are vaginal epithelial cells with many adherent *Gardnerella* morphotypes and wavy Gram-negative forms. However, there is no definitive laboratory test for BV and this microscopic appearance is only one of the criteria – it is consistent with the clinical diagnosis. Treatment is with metronidazole but recurrence is common. This condition is not sexually transmitted.

2 *Candidiasis ('thrush').* This is the second most common cause of vaginal discharge and is the most likely diagnosis in this case because of the history of recent antibiotic treatment. The yeast *Candida albicans* is part of the normal flora of the female genital tract, but it can overgrow if broad spectrum antibiotics kill off normal commensal bacteria. This leaves an 'ecological space' into which the *Candida* can expand. Thrush can also affect the oropharynx and oesophagus.

3 *Trichomonas vaginalis.* This is the third cause of vaginal discharge. It is much less common than the other two causes above and is sexually transmitted. It is a protozoan organism and sensitive to metronidazole.

The above are infections of the vaginal tract. *Chlamydia trachomatis* and *Neisseria gonorrhoeae* infections infect the cervix in the upper tract and can lead to pelvic inflammatory disease – note that chlamydia and gonorrhoea are often asymptomatic and may not cause vaginal discharge.

What do you think the most likely diagnosis is in this case?

Given the history of recent antibiotics and a white, itchy discharge ('*candida*' is Latin for 'white'), this patient's story is so typical of *Candida* infection that it would be reasonable to start treatment with an antifungal agent on the basis of a clinical diagnosis. Topical options include an azole, e.g. clotrimazole, or the polyene nystatin (see Antimicrobial chemotherapy, Part 1). Oral fluconazole has also been shown to be effective.

There are advantages to carrying out a gynaecological examination and sending a swab – low or high vaginal

Infectious Disease: Clinical Cases Uncovered. By H. McKenzie, R. Laing, A. Mackenzie, P. Molyneaux and A. Bal. Published 2009 by Blackwell Publishing. ISBN 978-1-4051-6891-5.

Table 13.1 Common causes of vaginal discharge.

Cause	Diagnosis	Comments
Bacterial vaginosis	Syndrome defined and diagnosed by clinical criteria (see text); typical microscopic picture supports diagnosis	Caused by a 'switch' of bacterial flora in the vaginal tract which is not fully understood. Symptoms may be mild and not associated with inflammation (hence 'vaginosis' and not 'vaginitis')
Candidiasis ('thrush')	Often made clinically, but microscopic appearance and culture of *Candida albicans* helps differentiate it from other causes	Itch is a common feature. Common after antibiotic treatment
Trichomonas vaginalis	Definitive diagnosis requires detection of the protozoan *T. vaginalis* by microscopy	Sexually transmitted

swab – to the lab for confirmation of your clinical diagnosis. Such swabs would be cultured for *Candida* and Gram stained to check for the presence of clue cells (see above) and help firm up the diagnosis. *T. vaginalis* is also identified by microscopy. Examination also allows detection of other anatomical or pathological abnormalities in the genital tract and, if appropriate, the opportunity to perform other investigations, e.g. a cervical smear.

You prescribe a 3-night course of clotrimazole pessaries. One week later your patient returns. The discharge is better, but she was impressed by your empathic consultation skills and would like to talk to you in confidence about sexually transmitted disease. She has read an article in a magazine about Chlamydia and realises that she is at risk because she has had several different sexual partners in the last year, without using any protection other than the oral contraceptive.

What are the potential complications of *Chlamydia* infection?

- Pelvic inflammatory disease.
- Ectopic pregnancy.
- Infertility.

How would you screen for it?

Most laboratories are now using nucleic acid amplification tests (see Laboratory diagnosis of infection, Part 1) and these work well on first void urine (FVU) as well as on endocervical swabs. Note that an FVU is the first 15–20 ml of urine flow and not necessarily an early morning urine – this is sometimes a source of confusion. Several studies have shown that self-collected vulval swabs are also an effective sample for diagnosis.

What other sexually transmitted diseases might you screen for?

- Gonorrhoea.
- HIV.
- Syphilis.
- Genital herpes.

Results

The laboratory report on the FVU states, 'Chlamydia DNA amplification test positive'.

What would you do now?

The patient should be treated with single-dose 1 g azithromycin. She should be encouraged to tell her recent sexual partners to visit their GP or a genitourinary clinic for testing. She should also be encouraged to use condoms in future to minimise the chances of further exposure to sexually transmitted infection.

CASE REVIEW

An 18-year-old student presented with itchy vaginal discharge 1 week after being prescribed co-amoxiclav for a presumed urinary tract infection. Vaginal candidiasis was the most likely diagnosis given the recent antibiotic history and the patient was treated empirically with clotrimazole pessaries. The condition resolved, but the patient returned to discuss her risk of *Chlamydia* infection. An FVU was tested by a nucleic acid amplification test and found to be positive. She was treated with a single dose of azithromycin and contact tracing was discussed.

KEY POINTS

- The commonest causes of vaginal discharge are bacterial vaginosis and *Candida albicans*. These are not generally due to sexual transmission.
- *Trichomonas vaginalis* is a less common cause of vaginal discharge and is sexually transmitted.
- *Chlamydia trachomatis* and *Neisseria gonorrhoeae* cause infection of the upper genital tract and are often asymptomatic in terms of vaginal discharge.
- Screening for sexually transmitted infection should be discussed opportunistically with patients who are at risk.
- Antibiotic therapy is a risk factor for fungal infections.
- Infection with *Chlamydia* can cause the long-term complications of pelvic inflammatory disease, ectopic pregnancy and infertility.
- Single-dose azithromycin is the antibiotic of choice for *Chlamydia* infection.

Further reading and information

SIGN Guideline 42. *Management of Genital Chlamydia Trachomatis Infection*. Available at http://www.sign.ac.uk/guidelines/fulltext/42/index.html.

Case 14 An outbreak of diarrhoea and vomiting on an orthopaedic ward

You are told by a nurse that a 75-year-old female patient has developed diarrhoea. She has passed two stools today and there was no blood, but they were described as 'watery'. She had a hip replacement 10 days ago and has been on co-amoxiclav for 7 days for a postoperative chest infection. On examination, she is in some discomfort with abdominal cramps, and there is localised tenderness and bowel sounds are present. Her pulse is 100 beats/min, BP 120/80 mmHg and her tongue is dry. She is in a four-bedded room.

What actions would you take?

• *Clinical assessment.* Rehydration should always be your first thought in the patient with diarrhoea, so if she is taking fluids by mouth, prescribe oral rehydration therapy; if clinically severe, set up a drip and give fluids intravenously. You would also want to check her urea and electrolytes to determine how dehydrated she is.

• *Infection control.* You do not know for certain that the cause of the diarrhoea is infective, but it would be sensible at this stage to arrange for the patient to be moved to a single room and barrier nursed pending laboratory results.

• *Establish cause.* You would want to send her stool for microbiology, ensuring that you request a test for *Clostridium difficile* toxin given the history of antibiotic treatment. Toxin detection, usually by enzyme immunoassay (EIA) (see Laboratory diagnosis of infection, Part 1) is the most useful marker of *C. difficile*-associated diarrhoea (CDAD), since culture takes several days and does not immediately distinguish between toxigenic and non-toxigenic strains.

Infectious Disease: Clinical Cases Uncovered. By H. McKenzie, R. Laing, A. Mackenzie, P. Molyneaux and A. Bal. Published 2009 by Blackwell Publishing. ISBN 978-1-4051-6891-5.

Whilst you are examining this patient, she tells you that the patient in the next bed also has diarrhoea. She is a 56-year-old woman with rheumatoid arthritis who had a knee replacement 5 days ago. You go to speak to her and find her reading a book and drinking tea. She admits to having diarrhoea but has no pain or discomfort. On further questioning she says that she has been 'a bit constipated' since the operation and she just happened to have some laxatives in her handbag, which have 'done the trick'. Her BP, pulse and abdominal examination are unremarkable. You explain to her the importance of only taking drugs given to her by the nurses and you inform the nurse in charge of the problem. On second thoughts, you ask for a stool to be sent to the laboratory if she has any more diarrhoea – just in case.

The following day, you have just started to clerk the routine admissions when the nurse in charge tracks you down. She tells you that the other two women in the four-bedded ward have been vomiting and one has developed diarrhoea. The nurse who originally told you about the first patient has phoned in sick, saying she has been up all night with diarrhoea and vomiting. You quickly examine the two 'new' patients and confirm the story of sudden onset projectile vomiting, accompanied by diarrhoea in one case. You find them miserable but not unduly ill. This is getting out of hand and you decide to phone the on-call microbiologist for advice. The microbiology registrar listens to the story and checks progress on the first patient's C. difficile toxin test. It transpires that it has just been done that morning and the result is positive – it is actually in his pile of results to be phoned.

What action would you take as a result of the positive toxin result?

You go back to examine the patient now confirmed as C. difficile toxin positive, who is now in a single room. She still complains of loose stools, a fact confirmed by the nurses. On examination she does not appear to be dehydrated.

Given that she has been on antibiotic for a postoperative chest infection, you examine her respiratory system but find her chest is clear.

What treatment do you initiate?

You stop the co-amoxiclav as it no longer seems indicated and it probably precipitated the CDAD. You start her on oral metronidazole and wash your hands carefully when leaving the room.

Relapse is common in *C. difficile* infection and repeat treatment may be required. The alternative to metronidazole for treatment is oral vancomycin – note that this is the only occasion when the oral form of vancomycin is used, since it is not absorbed.

How long does she have to stay in isolation?

The patient should remain in source isolation for 48 hours after she becomes symptomatic, i.e. she should remain there until her stools are formed. There is no clinical benefit in testing for 'clearance' of *C. difficile*. *C. difficile* is a spore-forming organism and contaminates the environment very easily. Spread of spores is much more likely when the patient is symptomatic.

Why is CDAD important?

CDAD is a common cause of hospital-acquired infection in hospitals. It is an anaerobic organism that can be present in the gastrointestinal (GI) tract of asymptomatic patients, but many strains have the potential to produce a toxin-mediated diarrhoea. This generally happens only when the normal flora of the GI tract has been altered by antibiotic treatment, allowing *C. difficile* to overgrow. Clindamycin treatment was originally identified as a major risk factor for CDAD, but it is now accepted that a wide range of broad spectrum antibiotics can precipitate the disease. The clinical spectrum of disease ranges from a mild antibiotic-associated diarrhoea to life-threatening pseudomembranous colitis. Recently, a highly virulent strain of *C. difficile* has been identified – the 027 ribotype – which is associated with more severe disease.

Since the organism is spore forming, spores contaminate the clinical environment and are difficult to eradicate. In addition, spores are not killed by alcohol hand rub, so you need to wash your hands with soap and water; the potential for patient to patient transfer of the organism by staff is considerable. Infection control measures are based on source isolation of patients, environmental cleaning and prudent antibiotic use.

What about the other patients with diarrhoea and vomiting?

By the time you have finished examining the patient with CDAD the infection control nurse arrives on the ward, alerted by the microbiology registrar to the two more recent diarrhoea and vomiting cases. The clinical picture strongly suggests norovirus infection and this is extremely infectious for both patients and staff. You hold an impromptu meeting with nursing staff and hear that the woman who took laxatives has now gone down with diarrhoea and vomiting. The infection control nurse suggests that this four-bedded bay should be closed to any further admissions and dedicated staff assigned to nursing duties in this bay.

Nursing a group of affected patients together in a limited area with dedicated staff is called 'cohorting'. Faeces samples from affected patients should be sent to the virology laboratory for norovirus testing. All staff should be reminded that they should not come to work if they develop diarrhoeal symptoms and/or vomiting and that they should remain off work for two full days after they become symptom free. Your local occupational health service should be alerted to staff involvement in this potential outbreak.

The following day the microbiology registrar phoned to tell you that the norovirus PCR is positive – but by this time, you are at home suffering identical symptoms to your patients. By the time you return to work 5 days later, the original C. difficile patient had emerged from isolation and was expecting to go home in a few days.

Why is norovirus important?

Outbreaks of diarrhoea and vomiting caused by norovirus have become a major public health problem, notably in hospitals, residential homes and cruise ships. The features of norovirus infection are outlined in Box 14.1.

Box 14.1 Common features of norovirus infection

Acute, sudden onset diarrhoea and vomiting (incubation time 15–50 hours)
Spontaneous resolution within 3 days
Low infective dose and high attack rate
Supportive treatment: fluids
Environmental decontamination of faeces/vomit vital to control outbreaks

CASE REVIEW

A 75-year-old woman admitted to an orthopaedic ward for a joint replacement was treated for chest infection postoperatively and developed diarrhoea, which was confirmed by the lab as EIA positive for *C. difficile* toxin. The patient was put into source isolation and treated successfully with metronidazole. Three other patients in the same four-bedded room and two members of staff developed vomiting and diarrhoea which was shown to be due to a different pathogen, norovirus. Infection control staff assisted with the management of the outbreak by cohorting these patients and reminding staff of optimal infection control measures.

KEY POINTS

- There are many possible causes of diarrhoea, not all of them infective, so each case requires investigation. Many alleged outbreaks turn out to be spurious, but you should always be alert to the possibility of case-to-case transmission.
- Isolation of individual patients or cohorting of groups of symptomatic patients in one ward or ward area can prevent spread.
- Adherence to infection control precautions by staff is essential – note that *Clostridium difficile* spores are not killed by alcohol rubs, so you need to wash your hands with soap and warm water.
- Microbiology and infection control staff can provide expert guidance in managing suspected outbreaks and should be informed of any potential outbreak.

Further reading and information

National *Clostridium difficile* Standards Group. Their report is available online on the Hospital Infection Society website, www.his.org.uk, in the Resource Library section.

Up-to-date information on *Clostridium difficile* and norovirus is available on the websites of the Health Protection Agency (www.hpa.org.uk) and Health Protection Scotland (www.hps.scot.nhs.uk).

Case 15 An antenatal visit

A 24-year-old woman attends the midwife-led antenatal clinic for her routine booking visit.

What infection screening tests should be recommended for all pregnant women in their first pregnancy?

- HIV serology.
- Syphilis antibody.
- Hepatitis B surface antigen.
- Rubella.

These routine serological tests are offered to this woman and she agrees to testing for HIV, hepatitis B, syphilis and rubella (no prior known testing or immunisation).

The results are back

As the duty doctor in the maternity ward you are alerted to the following results:

Hepatitis B surface antigen	*Positive*
Hepatitis B e-antigen	*Positive*
Hepatitis B e-antibody	*Negative*
Hepatitis B core IgM antibody	*Negative*
Hepatitis B core total antibody	*Positive*
Rubella IgG antibody	*Negative*
HIV antibody	*Negative*
Treponema pallidum antibody	*Negative*

Which of these results are clinically significant (requiring action either now or later)?

- The positive hepatitis B surface antigen and hepatitis B e-antigen results.
- The negative rubella antibody result.

Hepatitis serology can be difficult to interpret (review Table 4.1 in Case 4). The negative rubella antibody result shows that the woman has never had rubella and is therefore susceptible to the disease, with obvious risks to her fetus should she be exposed.

What should be done on the basis of these hepatitis B results?

The hepatitis B results show that this patient has chronic hepatitis B infection. There is a risk of vertical transmission to the baby, so action needs to be taken to minimise that. Arrangements should be made in advance of the expected date of delivery for hepatitis B immunisation of the baby in accordance with the Department of Health advice in *Immunisation against Infectious Disease 2006*, widely referred to as the 'Green Book'. Active immunisation for the baby involves four doses of hepatitis B vaccine: at birth, 4 weeks, 8 weeks and 1 year.

Is hepatitis B immunoglobulin indicated for the baby?

Yes: passive immunisation with hepatitis B immunoglobulin is indicated for babies of mothers who are e-antigen positive or who have no e-antibody or who have acute infection in pregnancy.

What is the implication for route of delivery and breastfeeding?

There is no implication for route of delivery. Breastfeeding is not contraindicated.

What other measures should be advised?

- Offer hepatitis B testing and immunisation to household and sexual contacts.
- Referral of the woman for assessment of liver disease.
- The woman should be offered testing for hepatitis C antibody.
- The woman should be offered testing for hepatitis A antibody, and, if negative, offered immunisation against hepatitis A.

Infectious Disease: Clinical Cases Uncovered. By H. McKenzie, R. Laing, A. Mackenzie, P. Molyneaux and A. Bal. Published 2009 by Blackwell Publishing. ISBN 978-1-4051-6891-5.

Why is hepatitis B screening done during pregnancy?

It is done so the baby can be protected once delivered. If a neonate becomes infected with hepatitis B, the risk of chronic hepatitis B infection is about 90%, with the associated risks of developing cirrhosis and hepatoma (primary liver cell cancer). A complete course of active immunisation prevents about 90% of these infections. The risk of transmission is far higher for babies born to a mother who is hepatitis B e-antigen positive; hence passive immunisation with hepatitis B immunoglobulin is also recommended for the babies of such mothers.

When is the greatest risk of mother to child transmission of hepatitis B?

About 95% of transmission occurs at or after delivery. It is therefore possible to build up protection to infection via immunisation during the long incubation time after exposure to hepatitis B. For the minority, around 5%, who have been infected *in utero*, immunisation is of no benefit but is not known to do any harm.

What follow-up should be arranged for the baby after completion of the course of immunisation?

The baby should be tested soon after 12 months for hepatitis B surface antigen to see whether or not chronic infection has been prevented. If infected, referral to a paediatric hepatologist is necessary. If uninfected, a pre-school booster dose of hepatitis B vaccine should be given.

> *The mother is offered and accepts hepatitis C antibody testing. The report received is:*
> *Hepatitis C antibody positive.*
> *Hepatitis C PCR positive.*

What do these results mean?

A positive hepatitis C antibody indicates infection at some time, but does not tell you if your patient has active infection. A positive PCR result means your patient is viraemic, i.e. has active infection.

What should be done on the basis of these results?

There is overlap with regard to hepatitis B.

- Testing for hepatitis C should be offered to household contacts.
- There is no active or passive immunisation available.
- There is no implication for route of delivery or breastfeeding.
- Refer for liver assessment, which can be after delivery.

What is the risk of transmission of hepatitis C to the baby?

This is about 5%. If the PCR had been negative, there would have been no risk to the baby. The baby should be tested, but the optimal timing for this is not yet clear. Two possibilities are antibody testing (once maternal antibody has gone, i.e. after 1 year) or PCR (at 2–3 months) followed by antibody testing if the PCR is negative.

What should be done on the basis of the negative rubella antibody result and why?

The result indicates susceptibility to rubella. Because of immunisation, rubella is now rare in the UK, but in the past it was a major cause of fetal infection with multisystem disease and congenital malformations if infection occurred in early pregnancy. Advice should therefore be given to report promptly any illness or contact with someone who might have rubella, i.e. any non-vesicular rash. Immunisation of the mother postpartum with MMR should be recommended (a live vaccine, so contraindicated during pregnancy). This should be either as two doses of MMR 1 month apart or one dose followed about 6 weeks later with testing to see if rubella antibody has developed. This therefore gives no protection for the current pregnancy, but will for any subsequent one – and also increases population immunity.

What would have been the implications if the HIV test had been positive following confirmatory testing?

- With no intervention, the risk of transmission to the baby is about 25%. Urgent referral is needed for assessment regarding antiviral therapy. This might be indicated immediately if the maternal CD4 count is low. However, if not, anti-retroviral treatment should be started in the second trimester and continued for the rest of the pregnancy.
- Genotypic resistance testing should be done on all new HIV diagnoses. About 10% of newly diagnosed cases in the UK show evidence of resistance to anti-HIV drugs.
- The aim of treatment is to decrease the maternal HIV viral load to an undetectable level.
- Offer testing to sexual contacts.
- Offer testing for hepatitis C.

Why is HIV screening recommended during pregnancy?

Measures taken during pregnancy, delivery and postnatally can decrease mother to child transmission from around 25% to under 2%.

When is the greatest risk of mother to child transmission of HIV?

Most transmission occurs during delivery, or via breast milk, but it can occur *in utero* (transplacental).

What is the implication for route of delivery and breastfeeding?

For many women, a planned caesarean section at 38–40 weeks is best and this is recommended for women on zidovudine monotherapy, or those who have a detectable viral load or hepatitis C co-infection. Normal vaginal delivery should be considered only for those with undetectable viral load. Invasive fetal monitoring and artificial rupture of membranes should be avoided. Breastfeeding is contraindicated where there is a safe and reliable alternative which can be continued until weaning. Mixing breastfeeding and formula feeding should be discouraged.

Why is syphilis screening recommended during pregnancy?

It is done so that maternal infection can be treated during pregnancy, thereby preventing infection of the fetus.

What is the risk of mother to child transmission of syphilis?

Congenital syphilis can occur if infection is transmitted from the mother to the baby *in utero*. Transmission usually occurs after the fourth month of gestation. Congenital syphilis may be latent or may manifest at birth. It can also lead to stillbirth or abortion.

In the first 4 years of infection in the mother, the risk of congenital syphilis is very high (approximately 70%).

How do we decide what screening should be done in pregnancy?

There are generic guidelines for screening. In outline, it is only worthwhile if a significant result leads to a clinical intervention that is of proven benefit – there is little point in screening for something if you cannot act on the result. Secondly, there has to be an assessment of the cost/benefit, something which can be difficult as it is not always easy to put a price on the value of preventing a disease (e.g. how do you measure the benefit of preventing one case of vertically transmitted HIV infection?). The detail of this is beyond the scope of this book, but you should be aware that there are generally agreed national policies based on such considerations, including a national policy for pregnancy screening.

For what other infections might there be a case for screening during pregnancy?

National policies vary depending upon the local incidence of particular infections, which may have a major impact on the cost/benefit of screening. The following are three other infections for which there may be screening during pregnancy.

- *Urinary tract infection.* Asymptomatic bacteriuria in pregnancy is associated with the later development of pyelonephritis and the risk of premature delivery. Studies show no evidence of benefit for antibiotic treatment of asymptomatic bacteriuria in the general population, but treatment is indicated in pregnancy. Screening by sampling a mid-stream urine towards the end of the first trimester is common.
- *Group B Streptococcus.* Screening of all pregnant women for genital carriage of this neonatal pathogen is performed in the USA. In the UK, the cost/benefit of screening is not so clear, as the incidence of neonatal infection is lower. Thus the Royal College of Obstetricians and Gynaecologists recommends intrapartum chemoprophylaxis with benzylpenicillin or clindamycin for certain high-risk groups, e.g. premature deliveries, prolonged rupture of membranes, fever in labour, a previous baby with group B streptccoccus infection or the incidental finding of a positive culture during the current pregnancy.
- *Toxoplasma gondii.* Serological screening is routinely offered in some countries, e.g. in France, where this condition is much more common. Pregnant women in the UK should be advised about the handling of cat litter and hygiene and should be tested if there is clinical suspicion of infection.

There are a number of other infections which should be considered in the investigation of pregnant women with relevant symptoms or risk factors (Table 15.1).

Table 15.1 Further infections to be considered in at-risk pregnant women.

Infection	Comments
Hepatitis C	Where there is a history of previous or current injecting drug use, hepatitis C antibody should be checked
Herpes simplex virus	Where there is symptomatic genital infection or visible lesions are present and there has been no previous testing
Varicella zoster virus	Where there is significant contact with chickenpox and/or shingles and there is no definite history of previous chickenpox or shingles
Parvovirus B19	Where there is contact with illness compatible with parvovirus infection
Chlamydia trachomatis or *Neisseria gonorrhoeae*	Both organisms can cause neonatal eye infections and *Chlamydia* can also cause neonatal respiratory tract infection
Listeria monocytogenes	*Listeria* causes a bacteraemic illness in pregnant women and is associated with stillbirth and neonatal meningitis. Unlike most bacteria, *Listeria* grows at 4°C and contaminates pâté and cheese made from unpasteurised milk

CASE REVIEW

A 24-year-old pregnant woman was found to have a positive test for hepatitis B surface antigen during routine antenatal screening. Further tests clarified that she had chronic hepatitis B infection. She was also positive for hepatitis C infection and negative for rubella and HIV antibodies. Active and passive immunisation was arranged for the baby at birth to reduce the chances of hepatitis B transmission. The mother was referred to a hepatologist for assessment of her liver disease and possible treatment of both hepatitis B and C. Postnatal MMR vaccine was recommended.

KEY POINTS

- Screening is worthwhile in pregnancy only if there is a worthwhile intervention to reduce maternal or neonatal morbidity.
- The risk of transmission of hepatitis B, HIV and syphilis from mother to baby can all be significantly reduced.
- The benefits of reducing the risk of hepatitis B, HIV and syphilis transmission are only available to those who have been tested.
- All pregnant women should be tested for hepatitis B, HIV and syphilis in every pregnancy.

Further reading and information

Department of Health. *Screening for Infectious Diseases in Pregnancy: standards to support the UK antenatal screening programme*. Department of Health, 2003, available at http://www.dh.gov.uk/en/Publicationsandstatistics/Publications/PublicationsPolicyAndGuidance/DH_4050934.

Department of Health. *Immunisation against Infectious Diseases 2006* (the 'Green Book'). Available at http://www.dh.gov.uk/en/Policyandguidance/Healthandsocialcaretopics/Greenbook/DH_4097254.

HIV in pregnancy information is available at www.bhiva.org.

Rashes and pregnancy guidance is available at http://www.hpa.org.uk/webw/HPAweb&Page&HPAwebAutoListName/Page/1200577813065?p=1200577813065.

Royal College of Obstetricians and Gynaecologists website has many guideline documents in this area: www.rcog.org.uk.

SIGN Guideline 92. *Management of Hepatitis C*. Available at www.sign.ac.uk.

Davidson SM *et al*. Perinatal hepatitis C infection: diagnosis and management. *Arch Dis Childhood* 2006; **91**: 781–5.

Case 16 Headache in a 35-year-old South African woman

A 35-year-old woman was admitted to hospital complaining of headaches. Her headaches began 3 weeks earlier and she had been aware of feeling hot and cold over the last 2 weeks. She had arrived in the UK from South Africa 6 months ago and was studying for a postgraduate degree.

What else would you like to know?

Headaches

- Are they worsening?
- Are there any aggravating or relieving factors?
- Where are they felt most?
- Are there any associated symptoms, e.g. vomiting or blurred vision?

The headaches were more or less constant and quite diffuse. She had no photophobia or vomiting. Simple analgesics such as paracetamol helped a bit and bending down or straining would make the headaches worse.

Other symptoms

- Had there been any systemic upset such as fevers or weight loss since the headaches began?
- Prior to the headaches had there been any other symptoms?

There was an awareness of feeling hot and cold for a couple of weeks that suggested fevers. No other symptoms were volunteered in relation to the headaches. However, she knew that she had been losing weight for the last 6 months, she thought around 10 kg. Her appetite was reduced and she had suffered some pain on swallowing, especially hot liquids.

Infectious Disease: Clinical Cases Uncovered. By H. McKenzie, R. Laing, A. Mackenzie, P. Molyneaux and A. Bal. Published 2009 by Blackwell Publishing. ISBN 978-1-4051-6891-5.

Past medical history

- Any previous history of headaches?
- Any problems that required medical attention in the past?

There was no past history of headaches. She had been married at the age of 24 and had become concerned when she had been unable to conceive after 5 years of marriage. She and her husband were assessed at a fertility clinic and she was informed that she had obstruction of the fallopian tubes. In vitro fertilisation had been suggested but her marriage broke up before this could be pursued.

What was found on examination?

The patient appeared lean (body mass index 19.5) with generalised lymphadenopathy. Oral candidiasis was evident (Plate 12). She had several raised, demarcated, purple lesions over the trunk, the largest being 2 cm in diameter. Fundoscopy revealed some haemorrhages and exudates inferotemporally in the right fundus. There was no papilloedema. She had an abnormal gait with a tendency to fall to the right and there was grade 4/5 right leg weakness with brisk reflexes in the right arm and leg.

What is the most likely explanation for this presentation?

There is really only one condition that is likely to predispose to all of the following:

- Headaches and right-sided weakness suggestive of an intracerebral space-occupying lesion.
- Oral thrush and odynophagia (pain on swallowing) suggestive of oesophageal candidiasis.
- Fundoscopic changes of a retinitis.
- Generalised lymphadenopathy.
- Unexplained skin lesions.
- Recent significant weight loss.

The patient is almost certain to have HIV infection with significant immune deficiency.

How do you confirm your suspicions?

Clearly an HIV test is the first step. Remember that it may take 3 months after HIV exposure for a patient to develop a positive antibody test. In this case you would expect the test to be positive as the patient has already developed symptoms suggestive of immune deficiency. Before asking a patient to have an HIV test it is important to explain what is involved and to discuss any concerns the patient may have, e.g. future work opportunities, having a family or the ability to obtain insurance or a mortgage. Up-to-date guidelines for HIV testing can be downloaded from the British HIV Association website (www.bhiva.org).

It is equally important to be clear about the benefits of knowing about a positive HIV test:

- To allow the timely introduction of antiviral therapy.
- To inform existing or future partners.
- To appropriately investigate symptoms in the context of HIV.
- To receive appropriate medical and social support.

The HIV test (an EIA, which can detect both HIV antibody and HIV antigen) was reported positive. A repeat sample was requested to ensure that it came from the correct patient (i.e. to exclude any patient or sample mix ups), and the result confirmed by Western Blot, a more sophisticated technique than basic EIA, which confirms the specificity of the antibody being measured.

What should be the next step?

As is often the case when patients present with symptomatic HIV infection, there are a number of problems that need to be sorted out:

- Oral and oesophageal candidiasis.
- Headache.
- Retinitis.
- Skin lesions.

How should these be prioritised?

The skin lesions are not likely to progress in the short term and the thrush can be treated empirically with antifungals, so the main areas of concern are the headaches (associated with focal neurology) and retinitis (which may be sight threatening if not treated).

What is the differential diagnosis for headaches in HIV disease?

The commonest causes would include:

- Cerebral toxoplasmosis.
- Central nervous system lymphoma.
- Cryptococcal meningitis.

Less common causes would include:

- Progressive multifocal leukoencephalopathy.
- Bacterial brain abscess including a tuberculous brain abscess.

What is the best way to sort out the cause of this patient's headaches?

Further examination will not help so imaging and other investigations should be arranged. A CT scan of the head is an obvious start.

The scan revealed lesions that enhanced with contrast (so-called 'ring enhancement') (Fig. 16.1). These are very typical of the abscesses caused by *Toxoplasma gondii*, i.e. this is cerebral toxoplasmosis. To confirm the diagnosis you would, if possible, perform a lumbar puncture and examine the CSF.

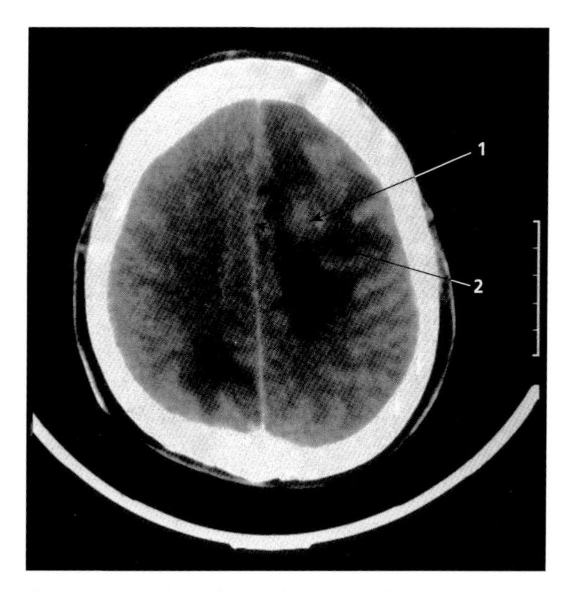

Figure 16.1 Ring lesion (arrow 1) suggestive of cerebral toxoplasmosis in a CT scan of the head. The darker area around the lesion is caused by cerebral oedema. Note the midline shift away from the side of the lesion (arrow 2).

PART 2: CASES

Results of CSF examination

	(Normal range)
White blood cell count	*15/μL (90% lymphocytes)*
Total protein	*0.8 g/L (0.15–0.45)*
PCR for T. gondii	*Positive*

These results confirm the diagnosis. The patient was started on treatment with sulphadiazine and pyrimethamine, which is a combination of antimicrobials effective against Toxoplasma. In addition, she received steroids to reduce the oedema that surrounded the abscesses. The CSF culture was later reported as negative for bacteria, including mycobacteria. Two weeks after starting treatment, the patient said that her headaches and weakness had resolved.

What about the retinitis?

The appearances of haemorrhages and exudates are very suggestive of cytomegalovirus (CMV) retinitis. The PCR for CMV in the blood is likely to be positive, but this can also be the case for people who have no evidence of CMV disease, so the diagnosis is principally a clinical one based on the appearance of the retina on fundoscopy. The patient needs to be treated as soon as possible with an antiviral drug that has activity against CMV. Without treatment the patient's vision could deteriorate dramatically within a matter of weeks.

She was started on valganciclovir – an oral antiviral drug that is active against CMV. The fundal changes were monitored and found to be improving after 4 weeks of treatment.

What about the HIV?

Broadly speaking, HIV can be asymptomatic (HIV infection) or symptomatic (HIV disease). Patients may be symptomatic from HIV itself (which can cause fevers, fatigue and weight loss) but they are often symptomatic because they have developed one or more opportunistic infections.

What are opportunistic infections?

An opportunistic infection is one that occurs in a host that has a compromised immune system – thus giving pathogens of relatively low virulence the opportunity to cause disease. Examples of opportunistic infections associated with HIV-related immunodeficiency are shown in Table 16.1.

Table 16.1 Examples of opportunistic infections associated with HIV disease.

Infection or organism	Typical clinical manifestation
Pneumocystis jirovecii	Pneumonia – symptoms of dry cough and breathlessness often develop over several weeks
Candida albicans	Oral or oesophageal thrush
Mycobacterium avium complex	Fevers, weight loss, anaemia – due to disseminated infection
Cytomegalovirus	Retinitis – visual field loss Encephalitis – fever and confusion Colitis – abdominal pain, fever and diarrhoea
Cryptosporidium parvum	Chronic watery diarrhoea
Toxoplasma gondii	Cerebral abscesses – headaches, fever and focal neurology
Cryptococcus neoformans	Chronic meningitis – headaches and fevers
JC virus	Progressive multifocal leukoencephalopathy

What about the patient's HIV infection?

This patient has suffered a number of these opportunistic infections (toxoplasmosis, oesophageal candidiasis, CMV retinitis) but treating these in a patient with HIV infection is not enough. In order to sustain improved health (and potentially restore normal health) the patient needs to have anti-HIV treatment that will lead, as far as possible, to normal immune function. This patient clearly has a deficient immune system, but to get some measure of this some more tests need to be carried out, as outlined in Table 16.2.

Laboratory assessment of HIV status

The results for this patient were:

	(Normal range)
CD4 count	*10×10^6 cells/L (430–1690)*
HIV viral load	*450 000 copies/ml*

These results are in keeping with severe immunodeficiency, as reflected in her opportunistic infections.

Table 16.2 Tests in HIV.

Type of test	Purpose of test
CD4 cell count (a measure of the number of lymphocytes that are CD4-receptor positive)	CD4 cells are destroyed by HIV and the risk of opportunistic infection or HIV-related tumours increases as the population of CD4 cells decline. The cell count predicts the risk of opportunistic infection and helps guide the need for antiviral therapy. CD4 counts of $<200 \times 10^6$/L are associated with a high risk of opportunistic infection
HIV viral load (a measure of the level of HIV in the blood)	This test is mainly used in patients receiving antiviral therapy. It tells you whether the drugs are working (in which case the HIV load should be undetectable)
HIV resistance testing (usually performed by RNA sequencing of the virus)	In patients with a detectable level of HIV in their blood it is possible to sequence the viral RNA to detect mutations that are known to be associated with drug resistance. This helps predict which antiviral treatments are most likely to be effective

In addition to treating her CMV and toxoplasmosis, the patient was started on co-trimoxazole as prophylaxis against Pneumocystis jirovecii pneumonia (formerly known as P. carinii pneumonia – often shortened to PCP). Patients with CD4 cell counts of less than 200×10^6 cells/L are at a greatly increased risk of PCP, but this risk can be dramatically reduced by prophylaxis with co-trimoxazole.

After the patient had been established on treatment for her opportunistic infections, she was started on a combination of HIV antiviral drugs. Two months later her HIV viral load was <40 copies/ml and her CD4 cell count had risen to 100×10^6 cells/L.

What now?

A CD4 cell count of 100×10^6 cells/L is still low and it is important to continue treatment for toxoplasmosis and CMV retinitis for the time being. If the immune system recovers sufficiently (CD4 cell count at least $>200 \times 10^6$/L on repeated measurements) it may be acceptable to change from a treatment dose to a prophylactic dose of drugs for toxoplasmosis and CMV. Eventually it may be possible to stop everything except the anti-HIV drugs that are maintaining the patient's immune function.

Do not forget about the skin lesions

These are typical of tumours known as Kaposi's sarcoma that commonly occur in advanced HIV infection and immunodeficiency. Kaposi's sarcoma is associated with a herpesvirus called Kaposi sarcoma-associated virus (human herpesvirus 8). Although the tumours occur most commonly in the skin, they can arise in deeper tissues such as the GI tract or the bronchi. These are very vascular tumours so those that occur in the gut can result in major GI haemorrhage and those occurring in bronchi can present with severe haemoptysis.

Widespread cutaneous or deep-seated Kaposi's sarcoma may be treated by chemotherapy, but this is not necessary for patients such as this one whose lesions were small and relatively few. As the patient's immune system recovered with antiviral therapy, the Kaposi's sarcoma lesions regressed until they were no longer visible.

What is the difference between HIV and AIDS?

The first definition of the acquired immune deficiency syndrome (AIDS) in 1982 was based on the diagnosis of certain conditions associated with severe immune deficiency, examples of which are listed above. Later definitions were extended to include patients with CD4 cell counts of less than 200×10^6 cells/L. Although the term AIDS is still in use, it is often better to think of patients with HIV as having asymptomatic or symptomatic disease. This patient presented with symptomatic HIV infection but, with antiviral therapy, she would be expected to become asymptomatic in time.

CASE REVIEW

A young South African immigrant was admitted with symptoms of headache, anorexia and weight loss. She was found to be HIV positive. Her CD4 T lymphocyte count was extremely low and her HIV viral load high, as a result of which she suffered from various opportunistic infections, namely CMV retinitis, cerebral toxoplasmosis and Kaposi's sarcoma. Specific treatments were started for the oropharyngeal candidiasis, CMV and *Toxoplasma* infections and she was commenced on prophylaxis for pneumocystis. Anti-HIV therapy was also commenced, and her various opportunistic infections resolved with the combination of specific treatment and the general improvement in her immune function. After 2 months' treatment, her HIV viral load was no longer detectable, but her CD4 count remained low and all treatments were therefore continued.

KEY POINTS

- The first line test for HIV infection is antibody/antigen measurement (serology). Patients should be counselled on the implications of testing and that there may be a window of around 3 months from the point of exposure until the test becomes positive.
- The progress of disease can be monitored by measuring the patient's CD4 count and HIV viral load.
- HIV infection causes the immune system to malfunction. The body thus loses the power to combat infections caused by microbes that are of very little consequence in people with fully functional immune systems. Such infections are called opportunistic infections.
- There are very specific guidelines about the management of opportunistic infections and management of HIV itself. This is a specialist area that requires expert advice or referral.

Further reading and information

British HIV Association provides regularly updated guidelines on many aspects of HIV care available at www.bhiva.org.

There are numerous good websites offering information on HIV, e.g. AIDS information for patients and healthcare workers available at www.aidsmap.com.

Case 17 A 64-year-old man with fever and rigors

A 64-year-old man was admitted to hospital with a 4-day history of fever, chills, rigors, sweating and malaise. He also complained of some right upper abdominal pain. He had returned from a 2-week cruise around the Caribbean 10 days before the onset of his symptoms.

What differential diagnosis would you think of at this stage?

1 Bacteraemia from an intra-abdominal focus.
2 Bacteraemia from a urinary tract infection.
3 Bacteraemia due to pneumonia.
4 Infective endocarditis.
5 Typhoid.
6 Malaria.

What details would you like to elicit from the history?

Associated symptoms

- Right upper abdominal pain: suggests either urinary or gallbladder disease, so it is important to enquire about the presence of jaundice, nausea, pale stool, dark urine and fatty food intolerance.
- Bloody diarrhoea: raises the possibility of recent amoebic dysentery, which can be complicated by the development of an amoebic liver abscess that could present in this way.
- Kidney infection: usually causes pain over the loin, but it may also be present in the right upper quadrant, so enquiring about urinary frequency and dysuria, haematuria and offensive urine is important.
- Pneumonia: may present in a non-specific way with relatively few respiratory symptoms, but enquire about cough, dyspnoea and chest pain.
- Malaria and typhoid: these can both present non-specifically, so it is important to clarify the risks by taking a detailed travel history.

Past medical history

Any past history of liver, gallstone disease or renal calculi or ureteric stricture should be sought as these problems may be recurrent.

Drug history

Enquire about recent antibiotic therapy – this could result in the condition being partly treated and result in a negative culture result.

Family history

Gallstone disease is often familial, as may be renal tract disease.

Further history is unhelpful, but in particular he has no ear, nose or throat, respiratory, bowel, skin or joint symptoms.

What are the key things you would look for on examination?

General examination

- State of alertness: confusion and diminished conscious level would suggest that the patient has a severe illness.
- Fever: a temperature of >38 °C or a low temperature of <35.5 °C would suggest the presence of infection (or inflammation). Initially normal temperature measurements, however, do not exclude the presence of serious infection.
- Tachycardia (pulse >100 beats/min) suggests a systemic inflammatory response.
- Hypotension (systolic BP <100 mmHg, diastolic BP <60 mmHg) is an important sign and may be due either to hypovolaemia (dehydration) or due to septic shock – see below.

Infectious Disease: Clinical Cases Uncovered. By H. McKenzie, R. Laing, A. Mackenzie, P. Molyneaux and A. Bal. Published 2009 by Blackwell Publishing. ISBN 978-1-4051-6891-5.

- Jaundice suggests biliary obstruction, either intrahepatic (e.g. liver abscess) or post-hepatic (e.g. gallstone obstructing the common bile duct, carcinoma of the head of the pancreas or cholangiocarcinoma). Jaundice may also be due to a pre-hepatic cause (e.g. haemolytic anaemia due to malaria). DIC may be due to overwhelming sepsis or a non-infectious cause such as lymphoma.
- Stigmata of endocarditis: splinter haemorrhages, finger clubbing, Janeway lesions and Roth spots. However, endocarditis usually presents subacutely after several weeks of illness rather than acutely, as in this man.
- Anaemia, which could be due to acute infection, haemolysis or blood loss.
- Bruising/bleeding suggests a bleeding tendency or thrombocytopenia which may be due to DIC.
- Evidence of weight loss: this may be due to subacute infection (e.g. endocarditis) or may suggest an underlying neoplasm or haematological malignancy.

Abdominal examination
There may be hepatomegaly, which can occur in typhoid, malaria, endocarditis or liver abscess. An area of tenderness between the ribs overlying the liver suggests a liver abscess. The gallbladder may be palpable or tender in cholecystitis. The spleen may be enlarged in endocarditis, malaria, typhoid or leishmaniasis. A tender kidney would suggest pyelonephritis, or a palpably enlarged kidney is likely to suggest polycystic disease, possibly renal carcinoma (a well recognised cause of FUO) or hydronephrosis. The presence of peritonism would imply a perforated bowel and the patient is likely to be complaining of abdominal pain.

Chest examination
Look for signs of a pneumonia, bronchiectasis or pleural effusion.

Cardiovascular examination
Check if there are any murmurs to suggest endocarditis.

Central nervous system examination
Look for focal neurological signs that may suggest a brain abscess or spinal epidural abscess.

Skin examination
Thorough examination of the skin is important to look for evidence of boils/abscesses or bed sores, particularly over the sacrum.

Joints examination
Examine for warmth, swelling, tenderness and reduced range of movement at any joint which could present a septic arthritis. Examine the spine for tenderness which may suggest vertebral osteomyelitis.

> *On examination he was pale and febrile with a temperature of 39.7°C, pulse 118 beats/min, BP 90/58 mmHg. No pallor, clubbing, jaundice, cyanosis or bruising was noted. There was tenderness in the right upper quadrant of the abdomen, but no organomegaly. Chest, cardiovascular system, central nervous system, skin and joints were all normal.*

What is your working diagnosis now?
Cholecystitis, ascending cholangitis or liver abscess.

What investigations would you like to perform in order to confirm the diagnosis?

Serum biochemistry
An elevated alkaline phosphatase and gamma glutamyl transferase, sometimes together with an elevated bilirubin, suggests the presence of biliary obstruction. This obstruction may be due to a gallstone obstructing the common bile duct, cholangiocarcinoma or carcinoma of the head of the pancreas.

Bilirubin is usually only elevated when the obstruction is complete. In the presence of a liver abscess, alanine aminotransferase is also elevated. The LDH may be abnormal in liver disease, but this enzyme is not organ specific, also being found in skeletal muscle, pericardium and red cells. Urea and creatinine may be abnormal in the presence of pre-renal renal dysfunction due to sepsis/dehydration or may be abnormal due to renal tract obstruction, e.g. hydronephrosis due to renal calculus.

Full blood count
In sepsis the haemoglobin may be low, normal or elevated (if severely dehydrated). White cell count is usually elevated (neutrophilia) in the presence of severe sepsis, but it may be low, which carries a worse prognosis. Platelet count is often reduced in the presence of sepsis, but if particularly low may reflect consumption coagulopathy (DIC).

C-reactive protein
CRP is typically elevated in bacterial infection.

Blood culture

If there is a suspicion of bacteraemia (e.g. rigors or subacute presentation are suggestive of endocarditis), two or three sets of blood cultures should be taken.

Urine microscopy and culture

This is the obvious test for urinary tract infection, but it may also be positive in a bacteraemic illness, e.g. typhoid fever.

Chest X-ray

A chest X-ray may show the presence of pneumonia or lung abscess.

Abdominal ultrasound scan

An abdominal ultrasound scan may show the presence of cholecystitis (with a thickened gallbladder wall, gallstones in the gallbladder, dilated common bile duct or intrahepatic bile duct due to distal obstruction or a liver abscess). If the abnormality is not seen or the anatomy is poorly defined, a CT scan would allow better definition. Magnetic resonance cholangiopancreatography (MRCP) is a useful, non-invasive imaging method for abnormalities in the biliary tree and pancreatic duct. Endoscopic retrograde cholangiopancreatography (ERCP) is another common method of imaging these ducts but is invasive (Fig. 17.1).

Ultrasound may also demonstrate a renal carcinoma, hydronephrosis or pus in the renal parenchyma or renal pelvis. An intravenous pyelogram would be needed to determine the cause of an obstructed ureter (e.g. calculus). Ultrasound may also detect abnormalities in the spleen, pancreas or pelvis.

Having organised all these investigations, you start intravenous saline in view of his hypotension and commence IV ceftriaxone and metronidazole because the history of rigors and his pyrexia of 39.7 °C suggest sepsis.

The following results are back

		(Normal range)
Haemoglobin	*14.1 g/dL*	*(13–18)*
White blood cell count	*17.2 × 10^9/L*	*(4–11)*
Platelet count	*189 × 10^9/L*	*(150–400)*
Urea, electrolytes and creatinine	*Normal*	
Serum albumin	*32 g/L*	*(37–49)*
Serum alanine aminotransferase	*60 U/L*	*(5–35)*
Serum alkaline phosphatase	*517 U/L*	*(45–105)*
Serum gamma glutamyl transferase	*602 U/L*	*(<50)*
Serum C-reactive protein	*350 mg/L*	*(<10)*

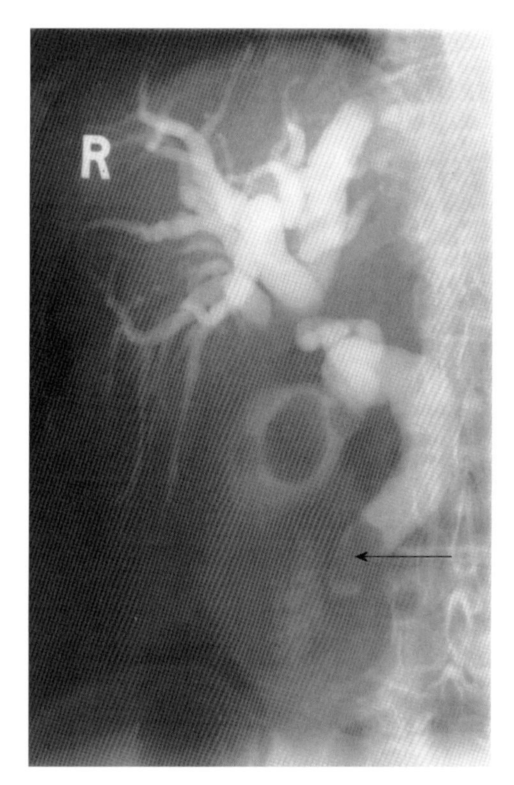

Figure 17.1 Gallstones evident on ERCP examination. Note the gallstone at the end of the common bile duct causing obstruction (arrow). Courtesy of Professor F. J. Gilbert.

Next day the abdominal ultrasound scan showed two small gallstones in the gallbladder (which was not inflamed). The common bile duct appeared normal. The liver, spleen, pancreas and kidneys were normal. The chest X-ray was normal. Urine microscopy and culture were negative. The patient's blood pressure had risen to 120/80 mmHg following infusion of 2 L saline over 2 hours, whilst his temperature had settled to 37.5 °C.

How do you interpret the results to date and what is your diagnosis now?

The biochemistry suggests biliary obstruction and, while the ultrasound shows no dilatation of the bile ducts, the presence of gallstones in the gallbladder suggests that the most likely diagnosis is ascending cholangitis. A gallstone may have already passed down the common bile duct. Features of sepsis with hypotension and the history of rigors suggest a likely bacteraemia secondary to biliary tract infection.

What is your approach to treatment?

The three main issues are resuscitation, antimicrobial therapy and treatment of the underlying structural problem.

Resuscitation

Hypotension (BP 90/50 mmHg) may be due to dehydration, severe sepsis or septic shock. The shocked, hypotensive patient should always be given intravenous fluids rapidly. Septic shock is defined as severe sepsis with persistent hypotension despite adequate fluid replacement, usually as judged by central venous pressure (5–10 cmH_2O). In this case, however, the patient's blood pressure responded to intravenous saline and he could therefore be regarded as having severe sepsis rather than septic shock (see definitions in Infection and immunity, Part 1).

Treatment of shock

If the patient had remained hypotensive, despite fluid resuscitation (with a normal central venous pressure), he would been regarded as having septic shock and transferred to the intensive care unit for further treatment and invasive monitoring. This treatment may have included intravenous adrenaline (which helps increase the blood pressure by increasing the total peripheral resistance) and intravenous noradrenaline (which increases the blood pressure by increasing the cardiac output). Organ failure requires support as required, e.g. mechanical ventilation for the respiratory system and renal dialysis for the renal system. Good control of blood sugar, possibly by using intravenous insulin, maintenance of a normal haemoglobin, adequate nutrition and oxygen, has been shown to improve the outcome of patients with septic shock and such interventions are routinely utilised.

Specific therapies for shock

There is an ongoing search for agents that interrupt the complex cytokine pathways which result in septic shock (see Infection and immunity, Part 1). Activated protein C is the only agent to date that appears to show some benefit when used appropriately. The clotting cascade is affected in severe sepsis with the most extreme example being DIC. Protein C inhibits thrombosis and promotes fibrinolysis, and levels have been shown to be low in septic shock. Therapeutic administration of activated protein C (drotrecogin) reduces the mortality in these patients but has the potential side effect of increased severe bleeding. In the UK it is recommended for treatment of adult patients with severe sepsis and multiple organ failure.

High-dose steroids have been shown to be associated with a worse outcome in patients with septic shock and so are not recommended. However, the use of low-dose (physiological) steroids has been more controversial with some reports suggesting a benefit.

Antibiotic therapy

Whether the patient has severe sepsis or septic shock, treatment with an appropriate antibiotic regimen that is likely to cover the infecting organism is essential. A broad spectrum regimen was required initially for this patient to treat the likely pathogens that cause biliary sepsis, e.g. coliforms (*Escherichia coli*, *Proteus mirabilis*, *Klebsiella oxytoca*) and anaerobes (*Bacteroides fragilis*). There are many possible combinations of antibiotics for this situation, but in general cover for coliforms and anaerobes is necessary. Appropriate treatment regimens would include ceftriaxone or cefotaxime or ciprofloxacin *plus* metronidazole; or co-amoxiclav *plus* gentamicin; or meropenem. Intravenous treatment is required initially and the choice of antibiotic must be reviewed in the light of any positive culture results.

Treatment of structural abnormality

It seems likely that this patient has already passed a gallstone, but if there was evidence that it had remained within the common bile duct, it might be necessary to perform an ERCP. During an ERCP, a sphincterotomy may be performed, which involves opening up the distal common bile duct to allow any remaining stones to be passed. Another technique is to recover the gallstone from within the common bile duct using a 'basket' device. If cholangitis or cholecystitis is a recurring problem, cholecystectomy may be indicated.

> *The following day the microbiologist telephoned to say that all four bottles from two sets of blood cultures had grown E. coli, provisionally resistant to amoxicillin but sensitive to ceftriaxone.*

Should you alter the therapy?

The patient had been treated with intravenous ceftriaxone and metronidazole since admission and was feeling a lot better. There was therefore no indication to change treatment.

He continued to make a full and uncomplicated recovery following a total of 1 week of treatment. The liver enzymes returned to normal and an outpatient MRI scan 4 weeks later showed gallstones in the gallbladder but no other abnormality.

CASE REVIEW

A 64-year-old man was admitted with right upper quadrant abdominal pain. On examination, he was found to be febrile and hypotensive. Abdominal ultrasound showed two gallstones in the gallbladder and blood cultures grew *E. coli*. Although the likely source of infection was the biliary tract, there was no evidence of any ongoing obstruction and no intervention (e.g. ERCP) was required. The patient was resuscitated with saline infusion and was treated with antibiotics to which the blood culture isolate was sensitive. He made a good recovery.

KEY POINTS

- There are a number of causes of fever with abdominal pain. Finding the exact cause requires a careful history, thorough examination and appropriate investigations.
- There are many causes of Gram-negative sepsis – urinary tract infection, pyelonephritis, diverticulitis, appendicitis, perforated colon, ischaemic colitis, pancreatitis, spontaneous bacterial peritonitis.
- Gram-negative sepsis originating from the abdomen is a life-threatening condition.
- Patients with severe sepsis or septic shock require intravenous fluids, oxygen, appropriate antibiotics and *senior help*!

Further reading and information

The Surviving Sepsis Campaign (SSC) is an initiative of the European Society of Intensive Care Medicine. Clinical guidelines and other information is available at www.survivingsepsis.org.

Case 18 A 42-year-old man with fever, cough and myalgia

A 42-year-old man was admitted to the acute medical admissions unit with a high fever, cough and myalgia. He had no significant past medical history and was not taking any regular medication.

What would you like to ask the patient?

- Duration of symptoms: acute illness or a prolonged history of fever and cough?
- Associated symptoms: sore throat, haemoptysis, weight loss or skin rash?
- Risk factors: smoking and/or alcohol abuse?
- Occupational history.
- Travel history.

Why is it important to elicit the above history?

Fever and cough are common presenting complaints in both community and hospital practice. A wide range of viruses and bacteria can both cause an acute illness characterised by the above symptoms (see Case 9). Fever, cough and weight loss over several weeks should always raise the suspicion of tuberculosis. Smoking can predispose to chest infections with or without a background of chronic obstructive pulmonary disease. Certain illnesses are often associated with particular occupations so the occupational history may be relevant. It is also very important to elicit a travel history as malaria can sometimes present with fever, myalgia and cough.

Your patient says that he was fit and healthy up until a couple of weeks ago. His illness started with what he describes as a 'chill' sensation, fever and headache. He took a day off work. Although not feeling completely well, he decided to go back to work but felt ill again 4–5 days later. He then consulted his general practitioner who arranged a hospital admission.

The patient stopped smoking 5 years ago. He consumes alcohol but only in moderation. He works in a meat factory. There is no history of haemoptysis or weight loss. He had been to Spain on a holiday with his family 6 months ago.

What should you do now?

Carry out a thorough examination and request some initial investigations: chest X-ray, full blood count, serum biochemistry, sputum and blood cultures.

The patient appears ill, his temperature is 38.5 °C, pulse 113 beats/min, BP 112/70 mmHg and respiratory rate 18 breaths/min. His eyes appear congested but there is no jaundice, rash or lymphadenopathy. There are a few crackles at the left base. Cardiovascular examination is normal. The spleen is just palpable and the liver is enlarged 2 cm below the costal margin.

What empirical treatment would you give pending laboratory results?

You should consider empirical treatment for chest infection. Oral amoxicillin-clavulanic acid plus clarithromycin is a good option. Amoxicillin-clavulanic acid may further derange the liver enzymes and you will need to keep a close watch on this patient's liver function.

The initial investigations are back

		(Normal range)
Haemoglobin	*14.6 g/dL*	*(13–18)*
White blood cell count	*9.1 × 10⁹/L*	*(4–11)*
Neutrophils	*5.6 × 10⁹/L*	*(1.5–7)*
Lymphocytes	*2.9 × 10⁹/L*	*(1.5–4)*
Platelets	*202 × 10⁹/L*	*(150–400)*

Infectious Disease: Clinical Cases Uncovered. By H. McKenzie, R. Laing, A. Mackenzie, P. Molyneaux and A. Bal. Published 2009 by Blackwell Publishing. ISBN 978-1-4051-6891-5.

Serum sodium	*135 mmol/L*	*(137–144)*
Serum potassium	*4.7 mmol/L*	*(3.5–4.9)*
Serum urea	*9 mmol/L*	*(2.5–7.5)*
Serum creatinine	*168 μmol/L*	*(60–110)*
Serum total bilirubin	*20 μmol/L*	*(1–22)*
Serum alanine aminotransferase	*166 U/L*	*(5–35)*
Serum alkaline phosphatase	*134 U/L*	*(45–105)*
Serum gamma glutamyl transferase	*196 U/L*	*(<50)*
Serum C-reactive protein	*88 mg/L*	*(<10)*
Chest X-ray	*Normal*	
Sputum	*Scanty polymorphs and no organisms*	
Culture results are awaited		

What is your differential diagnosis?

1 Viral illness, e.g. influenza.

2 Lower respiratory tract infection.

3 Given that he works in a meat factory, zoonotic infections must also be considered, e.g. Q fever, leptospirosis, brucellosis.

A day later, the sputum culture showed only normal upper respiratory tract flora.

What other tests would you carry out?

Blood cultures have already been sent, but further sets could be useful. Viral hepatitis serology is indicated given his elevated liver enzymes. As the patient works in a meat factory, you should also send a serum sample for serology for Q fever (*Coxiella burnetii*), brucellosis and leptospirosis, all of which are zoonotic infections. It is best to consult the microbiologist or infection specialist before requesting these tests as the laboratory can help you in several ways if they know the patient's history. For example, the blood cultures could be incubated for a prolonged period (say 2 weeks instead of the usual 5 days) for successful isolation of *Brucella*, which is a fastidious (slowly growing) organism.

The next morning, you phone the laboratory regarding the results of the serological tests for Q fever, brucellosis and leptospirosis. The enzyme immunoassay test for leptospira serology is equivocal. The titres for Q fever and brucellosis are awaited and the hepatitis serology is negative.

What is your diagnosis now?

This meat factory worker presented with fever, cough, myalgia and red eyes. The clinical picture, taken together with the abnormal liver enzymes, renal dysfunction and leptospira serology result are consistent with a diagnosis of leptospirosis. In fact, symptoms of a 'viral' illness are a common presenting feature of this infection. The leptospiral serology result could represent a false positive but it is hard to ignore under the circumstances. Some patients with leptospirosis present with jaundice, but it seldom causes a marked increase in the liver enzymes, such as that commonly seen in viral hepatitis.

The patient's case is discussed with an infectious diseases specialist and microbiologists. Q fever and brucellosis serology has now been reported as negative. The consensus is to treat the patient for leptospirosis. The antibiotic of choice is benzylpenicillin. Leptospirosis is notifiable so you should also inform the local health protection team.

How do you confirm a diagnosis of leptospirosis?

This is not difficult but it takes some time. You need to send a convalescent serum sample taken 2–3 weeks after the acute phase sample. A fourfold rise in antibody titre is diagnostic and this would normally be confirmed by a reference laboratory. Alternatively, if the acute phase serum was negative, a seroconversion (i.e. acute phase serum negative but convalescent serum positive) is taken as proof of infection (see Laboratory diagnosis of infection, Part 1).

Leptospira can be cultured but this is not straightforward and culture methods are not widely available.

What is leptospirosis and how did your patient acquire this infection?

Leptospira are spirochaetes related to *Treponema* and *Borrelia*. There are two species – *L. interrogans* and *L. biflexa* – and each has a number of serovars; more than 200 serovars have been described. Determination of serovars is important for epidemiological purposes.

Animals such as rats, pigs, dogs and horses are reservoirs of leptospira and can shed large numbers of the organisms in urine, even when asymptomatic. Leptospira enter the body through small cuts and abrasions on the skin or via the intact mucous membrane. It is quite likely that your patient sustained a small cut at work which was the portal of entry for the leptospira. Leptospirosis is endemic in tropical countries of Asia but infections occur worldwide.

Infection acquired as a result of exposure to animals is called zoonosis.

What happened next?

The patient gradually improved on treatment with intravenous benzylpenicillin. On day 7 he was discharged from the hospital. He made a complete recovery.

Three weeks later, a convalescent serum sample sent by his GP was forwarded to the reference laboratory and showed an elevated Lepstospira antibody titre of 320 (serovar L. hardjo) – an eightfold rise from the acute sample titre of 40. The diagnosis was therefore confirmed.

CASE REVIEW

A 42-year-old meat factory worker was admitted to hospital with a 2-week history of fever, cough and myalgia. On examination, his spleen was just palpable and his liver was enlarged just below the costal margin. Biochemistry revealed a moderate increase in liver enzymes and decreasing renal function. The differential diagnosis focused initially on chest infection and appropriate antibiotic treatment was started for this, but following a clear chest X-ray and negative sputum culture, the diagnosis was re-considered. Zoonoses relevant to his occupational history were considered and serology for leptospirosis was weakly positive. This diagnosis was consistent with the clinical findings, so benzylpenicillin was started. The patient made a good recovery and follow-up serology confirmed the diagnosis by virtue of a rising titre of antibody.

KEY POINTS

- Infections can often present with non-specific symptoms that do not point to a particular diagnosis.
- The occupational history may be crucial, as it was in this patient.
- Leptospirosis often presents as a flu-like illness with abnormal liver and kidney function. It responds well to treatment with penicillin.

Case 19 A 75-year-old man with a sore hip

A 75-year-old man is admitted with a 4-week history of pain around his left hip joint. He had a prosthetic hip placed on the same side 5 months prior to his current admission.

What are the likely causes of pain around a prosthetic joint?

Infection in the prosthetic hip and mechanical loosening of the prosthesis are the commonest causes of joint pain in someone with a prosthetic joint. It is sometimes difficult to differentiate the two conditions and you should take a thorough history and undertake a careful examination to help do this. You must also find out whether your patient has predisposing conditions, e.g. rheumatoid arthritis, diabetes mellitus and steroid therapy, which all increase the risk of prosthetic joint infection.

Look for the common signs of infection

- Is the patient febrile?
- Are there any other systemic signs of infection?
- Is there pus leaking from the surgical wound site?

Continuous pain suggests an infective process, while mechanical problems cause pain only on motion. Fever, hypotension and raised inflammatory markers suggest an ongoing infection. Pus leaking from the wound site is highly suggestive of infection.

Your patient is afebrile and shows no systemic signs of infection. He is in constant pain that worsens on movement. He describes the pain as a dull, aching sensation. The surgical site appears well healed as expected at 5 months following the procedure. He does not have rheumatoid arthritis and he is not a known diabetic. His joint was replaced for a chronic debilitating osteoarthritis.

Infectious Disease: Clinical Cases Uncovered. By H. McKenzie, R. Laing, A. Mackenzie, P. Molyneaux and A. Bal. Published 2009 by Blackwell Publishing. ISBN 978-1-4051-6891-5.

Does he have an infection of his prosthetic hip?

While fever is a common symptom of an infective process, absence of fever does not rule out infection in a prosthetic joint. In fact, only half of the patients with prosthetic joint infections have fever due to the localising nature of the infective process. Similarly, the superficial wound may have healed completely by the time an infection is detected in the deep joint tissue and prosthesis. This is because the bacteria that are usually associated with infection in foreign materials *in situ* are often of low virulence. Such an infection is slow to progress but can be debilitating.

What investigations would you like to carry out?

Radiological investigations are required to confirm whether the prosthesis has loosened. A simple X-ray can reveal joint fractures and establish the location of the prosthetic head of femur in relation to the acetabular cup. A bone scan, CT scan, MRI scan and arthrography may also be of help. Ideally, aspiration of the joint with culture of the fluid should be performed to confirm the presence of infection, identify the organism and determine its antibiotic susceptibility.

Your patient had a CT scan that showed evidence of infection in his hip.

How do you manage this case?

There are several ways to treat prosthetic joint infections. If patients present very early during the course of infection (within 1 month of onset), it is possible to treat with simple drainage and antibiotics. The prosthesis is not always removed in such cases. Infections that present late almost always require prosthesis removal. This can be accomplished with the help of a single- or two-step procedure. In a single-step operation, the infected prosthesis is removed and a new one placed in at the same time. In a two-stage operation, the infected prosthesis is removed and antibiotics are administered for several weeks before a new prosthesis is placed.

Factors that determine management include the duration of symptoms, the pathogen causing the infection and the fitness of the patient for repeat surgery.

Prosthetic material *in situ* gets infected in a variety of ways. Any episode of bacteraemia can seed foreign material with bacteria. Such haematogenous spread can occur early in the postoperative phase. For example, a central vascular catheter-associated bacteraemia caused by *Staphylococcus aureus* can potentially seed a prosthetic joint. More commonly, the infection is introduced locally either directly during the operative procedure or indirectly following infection of the operative wound. Coagulase-negative staphylococci are the commonest bacteria associated with prosthetic joint infections.

It was decided that your patient should undergo a two-stage procedure. The orthopaedic surgeon removed the prosthesis and multiple tissue samples and swabs taken intraoperatively were sent to the microbiology laboratory for culture. During this first stage, cement spacers were placed within the joint cavity. All tissue samples grew coagulase-negative staphylococci sensitive to vancomycin, teicoplanin, gentamicin and rifampicin.

Which antibiotic would you use for treatment?

You prescribe teicoplanin as glycopeptides (e.g. vancomycin or teicoplanin) are the drugs of choice for treating deep-seated infection with coagulase-negative staphylococci. You tell him he will need a full 6-week course of intravenous treatment.

Successful treatment of infections requires the use of appropriate antibiotics. The choice of agent depends upon the site of infection, the causative organism, and the pharmacokinetics of the antibiotic and its toxicity. This is particularly important when infection involves the heart valves, joints, bones and meninges. Antibiotics should be able to penetrate into the deep tissues, should ideally be bactericidal, and should be well tolerated by patients. Infections associated with foreign materials such as the prosthetic joints, prosthetic heart valves and long-term vascular catheters require aggressive management. Bacteria causing infection at these sites form a biofilm. Bacterial cells inside a biofilm attach themselves to each other and also to the underlying tissue matrix. Biofilm is therefore like a protective cover inside which bacteria can thrive, given the fact that immune cells and antibiotics often fail to penetrate the film. Thus, antibiotics that are used to treat such infections should not only be effective against the infective strains, but should also be able to penetrate the biofilms in order to affect cure.

Apart from a two-stage operation, management of prosthetic joint infection often requires antibiotic treatment for a period of 4–6 weeks. Coagulase-negative staphylococci lead to a slow, indolent infection. These staphylococci commonly reside on the surface of the skin. Consequently, they are often contaminants in blood cultures that are drawn without adequate sterile precautions. However, when isolated from multiple specimens, they should not be dismissed as contaminants without a thorough investigation. Coagulase-negative staphylococci are usually resistant to multiple antibiotics. In fact, their antibiotic susceptibility pattern is often similar to that of methicillin-resistant *S. aureus* (MRSA). Like MRSA, they are resistant to commonly used antibiotics such as benzylpenicillin, β-lactamase-stable penicillins (e.g. flucloxacillin), macrolides (e.g. erythromycin), lincosamides (e.g. clindamycin) and quinolones (e.g. ciprofloxacin).

Glycopeptides are the drugs of choice for treating deep-seated infection with coagulase-negative staphylococci and MRSA. The advantage of teicoplanin is that after an initial loading dose, it can be given once daily. This makes it an attractive option for 'OPAT'. A second antibiotic, such as rifampicin or fusidic acid, is sometimes added to the regimen but there is currently no evidence base to support this.

What is OPAT?

OPAT stands for out-patient antibiotic therapy. It is not desirable for a patient to be in hospital only in order to receive parenteral antibiotics if he or she is otherwise well and able to manage at home. Parenteral medications can be administered by a carer or, in some cases, patients themselves can deliver the medication via a peripherally inserted central catheter.

You could consider your patient for OPAT and discuss this with your team.

What happened next?

Your patient completed a 6-week course of teicoplanin, mostly outwith the hospital. Surgery to insert a new prosthesis was uneventful and swabs taken during the operation were subsequently reported as 'no growth'.

CASE REVIEW

A 75-year-old man presented with hip pain 5 months after a hip replacement operation. A CT scan confirmed infection and a two-stage procedure was undertaken, the joint being removed in the first stage and multiple samples sent to the lab for culture. Coagulase-negative staphylococci were grown from all samples and the patient was treated with IV teicoplanin for 6 weeks at home. At the second-stage operation, swabs from the operative site were taken and were negative on culture. A second prosthesis was inserted successfully.

KEY POINTS

- Foreign bodies of any kind, e.g. prosthetic joints or valves, are an infection risk.
- Organisms that infect foreign bodies such as prostheses are often of low virulence but are good at forming biofilms on foreign material, e.g. coagulase-negative staphylococci.
- In patients with suspected prosthetic device infection, review their medical records for evidence of MRSA infection or colonisation, as this is likely to be the cause of the infection.
- Removal of the prosthesis is often the only way to cure the infection permanently.
- Infected prosthetic hips are often treated by a two-stage procedure, with removal of the infected joint followed by antibiotic treatment and then subsequent joint replacement.

Further reading

Tobin EH. Prosthetic joint infections: controversies and clues. *Lancet* 1999; **353**: 770–771.

Case 20 A 24-year-old man with acute myelogenous leukaemia who has developed a fever

A 24-year-old man with acute myelogenous leukaemia (AML) is undergoing his remission–induction chemotherapy cycle with cytarabine and doxorubicin. On the 8th day following chemotherapy, the patient is febrile. His temperature is 39.2 °C. Some of his haematology and biochemical parameters are as follows.

		(Normal range)
Haemoglobin	*9.8 g/dL*	*(13–18)*
Red blood cell count	*2.1×10^{12}/L*	*(4.3–5.9)*
White blood cell count	*0.3×10^9/L*	*(4–11)*
Neutrophil count	*0.2×10^9/L*	*(1.5–7)*
Eosinophil count	*0.0×10^9/L*	*(0.04–0.4)*
Lymphocyte count	*0.1×10^9/L*	*(1.5–4)*
Monocyte count	*0.0×10^9/L*	*(<0.8)*
Platelets	*22×10^9/L*	*(150–400)*
Serum sodium	*134 mmol/L*	*(137–144)*
Serum potassium	*4 mmol/L*	*(3.5–4.9)*
Serum chloride	*102 mmol/L*	*(95–107)*
Serum urea	*7 mmol/L*	*(2.5–7.5)*
Serum creatinine	*96 µmol/L*	*(60–110)*
Serum total bilirubin	*12 µmol/L*	*(1–22)*
Serum alanine aminotransferase	*26 U/L*	*(5–35)*
Serum alkaline phosphatase	*197 U/L*	*(45–105)*
Serum C-reactive protein	*53 mg/L*	*(<10)*

How should this patient be managed?

Patients with haematological malignancy are treated with cytoreductive chemotherapy that kills the malignant cells. Unfortunately, these chemotherapeutic agents have a narrow therapeutic window. Thus, they have a major effect on the normal tissues including the cells of the bone marrow. A potential consequence of this effect is pancytopaenia. The neutrophil and lymphocyte counts in these patients can become so low as to be undetectable. Neutrophils are critical in host defence mechanism against various infections.

The host is therefore left vulnerable to serious bacterial, viral and fungal infections. Such infections can either be endogenous or exogenous.

- Endogenous infections are caused by organisms that are commensals in the human body in health. Chemotherapy damages the mucosal lining, thus making translocation of normal flora into the blood stream relatively easy. In this manner, streptococci breach the damaged oral mucosa leading to bacteraemia. Similarly, coliforms and *Candida* can cause a blood stream infection following damage to the intestinal mucosa. Vascular catheters are a ready source of infection with coagulase-negative staphylococci that are normal commensals of the skin.
- Exogenous infections on the other hand are acquired following exposure to an infectious agent outside the host.

While all attempts must be made to establish the source and aetiology of infection, it is imperative to start appropriate antimicrobial agents in patients with febrile neutropaenia without waiting for laboratory results. The choice of these agents depends upon various factors, which include:

- The nature of the underlying disease.
- The type of cytoreductive therapy.
- Common causes of infection at various time points following chemotherapy.
- Any past history of infections such as invasive fungal infections prior to the current chemotherapy cycle.
- Any antimicrobial prophylaxis that the patient might be taking.
- Local antibiotic susceptibility data.

Infectious Disease: Clinical Cases Uncovered. By H. McKenzie, R. Laing, A. Mackenzie, P. Molyneaux and A. Bal. Published 2009 by Blackwell Publishing. ISBN 978-1-4051-6891-5.

What should be the immediate line of management?

You must take a thorough history and perform a systemic examination to find the cause of the fever. Infection should always be suspected during the evaluation of a patient with febrile neutropaenia. Infection could arise from the damaged mucosal sites as described above or could be associated with a vascular device such as a Hickman line. It could also be associated with a specific system or organs such as the lungs, the central nervous system, the gut and the genitourinary tract.

Neutropaenic patients are prone to develop pneumonia. Does your patient have cough with shortness of breath? A chest X-ray should be performed to look for features of a respiratory infection.

Take blood cultures from the ports of any venous catheter and also from a peripheral vein. Positive culture results will influence the antibiotic treatment later. Sputum and urine should also be cultured.

What antibiotics would you like to start?

Most haematology units will have an empirical antibiotic policy. You should refer to your local policy, but it is important to understand the principle behind the choice of agents. Antibiotic selection depends upon the organisms that you wish to target. Antibiotics should always cover Gram-negative organisms because these bacteria are associated with high mortality rates, approaching 40%. Traditionally, combination therapy with a broad spectrum β-lactam antibiotic and an aminoglycoside has been the preferred treatment. Piperacillin-tazobactam and gentamicin is a popular choice but there are several other options available. It is generally agreed that exclusive cover for Gram-positive bacteria with agents such as vancomycin is not required in the early stages, but policies differ between institutions.

How should you follow up the patient?

Haematology patients on chemotherapy require close monitoring. You should do a daily (and sometimes even more frequent) full blood count and biochemistry. It is important to monitor the total and differential white cell count and renal function. Chemotherapeutic drugs have wide ranging tissue toxicities. For example, you should monitor serum uric acid levels with cytarabine therapy.

How is your patient doing?

Forty-eight hours after starting piperacillin-tazobactam and gentamicin, the patient continues to be febrile. Microbiology results have not been positive so far. On the ward round, the possibility of adding specific Gram-positive cover was discussed and intravenous teicoplanin was added to the empirical regimen.

Over the next 48 hours, your patient complains of a dry cough and a vague chest discomfort. You auscultate his chest but no crackles or wheezes are heard and your consultant asks you to explore the possibility of carrying out further investigations to rule out a chest infection. A repeat chest X-ray shows an infiltrate. The blood cultures taken at the time when the patient first developed fever did not yield any organisms.

A further investigation to consider

You should consider getting a CT scan of the chest in the light of the ongoing symptoms. Whilst a CT scan cannot provide a definitive microbiological diagnosis, this procedure can provide evidence for some aetiologies. For example, focal lesions are commonly seen with fungal infections, while bacterial infections cause segmental lesions. A 'halo sign' on a CT scan suggests pulmonary aspergillosis. Diffuse bilateral infiltrates should suggest a viral aetiology.

Pneumonia is usually seen late in the course of neutropenia, but can occur at any time. You should note that CT scan findings can be entirely normal early during the phase of pneumonia. A negative CT scan does not rule out an infection.

The CT scan result shows an area of consolidation surrounded by a typical 'halo', consistent with pulmonary aspergillosis.

Should therapy be modified in the light of the above finding?

This is strong evidence to support the addition of an agent that is active against *Aspergillus*. This needs to be discussed with a multidisciplinary team including the haematologists, microbiologists and pharmacists. Several antifungal agents are active against *Aspergillus*. These include amphotericin B, liposomal amphotericin and voriconazole. Unfortunately, these agents also have serious side effects. Treating fungal infections in neutropaenic patients is a very specialised area.

What additional tests can be done to firm up the diagnosis?

You should also consider further investigations such as:

• *Fibreoptic bronchoscopy.* Bronchoscopy with lavage is a useful diagnostic procedure. This will provide a high quality sample from the lower respiratory tract that can be subjected to a variety of microbiological tests. These include routine bacterial and fungal cultures (e.g. to confirm the presence of an *Aspergillus* species), culture for *Mycobacteria* and *Nocardia*, and a PCR for *Pneumocystis*. A bronchial biopsy or an open lung biopsy may also be diagnostic of *Aspergillus* infection. These are highly invasive techniques and patients may not always be fit to undergo them.

• *Serum antigen test for Aspergillus.* This may sometimes be useful in confirming invasive disease. The test detects *Aspergillus* galactomannan antigen. It is very rare for *Aspergillus* to grow from blood cultures.

• Urine antigen test for *Legionella*. This is a simple non-invasive test that detects the presence of *Legionella* antigen in urine.

What happened next?

Your patient had a low platelet count and bronchoscopy was not undertaken because of the risks associated with thrombocytopaenia. The serum galactomannan antigen test was negative and so was the urine antigen test for Legionella. He was started on voriconazole. His neutrophil count gradually improved and a repeat chest X-ray showed improvement. He was eventually discharged on consolidation therapy for his AML.

CASE REVIEW

A 24-year-old man with acute myelogenous leukaemia developed a fever while neutropaenic following chemotherapy. He was treated according to local prococols for febrile neutropaenia with intravenous piperacillin-tazobactam and gentamicin. Teicoplanin was added for Gram-positive cover when there was no response in 48 hours. His chest CT scan was suggestive of pulmonary aspergillosis for which he received voriconazole. It was not possible to confirm the diagnosis of pulmonary aspergillosis microbiologically as the patient was not well enough for invasive sampling (i.e. a BAL). His condition resolved slowly and he was discharged on consolidation treatment.

KEY POINTS

- Febrile neutropaenia requires immediate action. All possible samples should be sent to the lab for culture and the patient commenced on initial antibiotic cover according to local guidelines. This is usually directed at Gram negatives (e.g. piperacillin-tazobactam and gentamicin).
- Gram-positive cover (e.g. teicoplanin or vancomycin) is added after 48 hours or so if there is no improvement or if there are no laboratory results to inform management.
- If there is no response, antifungal cover is added. Several new agents are available and specialist advice is helpful at this stage.
- Haematology is a specialised field and a multidisciplinary approach is essential for effective management of these patients.
- CT scans may be useful in the diagnosis of disseminated infection.

Further reading

Hughes WT *et al.* 2002 Guidelines for the use of antimicrobial agents in neutropaenic patients with cancer. *Clinical Infectious Diseases* 2002; **34**: 730–751.

British Committee for Standards in Haematology. Guidelines on the management of invasive fungal infection during therapy for haematological malignancy. Available at www.bcshguidelines.com.

Case 21 A 53-year-old man with fever, severe back pain and abdominal pain

A 53-year-old man was admitted on a Sunday evening to the acute medical assessment unit. On the Monday afternoon, you get a handover from the weekend medical team. It has been an unusually busy weekend and you are told that this patient requires further review. You are told that this is his third admission with urinary tract infection (UTI) in the last 5 months. His admission records show that the previous two episodes were treated with intravenous co-amoxiclav. During the current admission, he has been treated with intravenous ciprofloxacin pending investigations.

What will your first line of management be?

The immediate steps that you should take would be:

- Review the history and re-examine the patient.
- Follow up any investigations that have been done over the weekend.

History and examination

Your patient was admitted last night with a history of high fever and severe back pain and abdominal pain. The onset of pain was sudden, though he had some vague abdominal discomfort for almost a week prior to admission. He vomited soon after the onset of pain. His wife phoned the out-of-hours GP service and an admission was arranged. The patient has no history of diabetes, hypertension or cardiac or respiratory disease. He is a non-smoker and consumes alcohol in moderation.

On admission, his temperature was 40.2 °C, BP 100/88 mmHg, pulse 102 beats/min and respiratory rate 15 breaths/min. The night team did the initial investigations including full blood count, serum biochemistry, urinalysis, blood culture and urine culture. He was then started on ciprofloxacin and analgesia and catheterised.

Your patient looks brighter today. His temperature is now 38 °C. He complains of pain in his left flank and he also has generalised abdominal pain. On examination, the right flank and epigastric region are tender.

He has passed only 30 ml of urine since he was catheterised. Urinalysis showed leucocytes +++, nitrites ++, proteins +++ and blood ++. The pH of the urine was found to be 8.7.

Results of biochemical and haematological investigations on admission are as follows:

		(Normal range)
Haemoglobin	12.3 g/dL	(13–18)
Red blood cell count	3.7×10^{12}/L	(4.3–5.9)
White blood cell count	20.3×10^{9}/L	(4–11)
Platelet count	152×10^{9}/L	(150–400)
Serum sodium	143 mmol/L	(137–144)
Serum potassium	4.7 mmol/L	(3.5–4.9)
Serum chloride	104 mmol/L	(95–107)
Serum bicarbonate	31 mmol/L	(20–28)
Serum calcium	2.4 mmol/L	(2.2–2.6)
Serum urate	0.302 mmol/L	(0.23–0.46)
Serum urea	13 mmol/L	(2.5–7.5)
Serum creatinine	196 µmol/L	(60–110)
Serum C-reactive protein	353 mg/L	(<10)

The microbiologist phones to say that the blood cultures taken yesterday have Gram-negative bacilli in each bottle. The urine culture has grown 10^5/ml of a coliform organism. Sensitivities and identification of these isolates should be available tomorrow.

A provisional diagnosis of acute pyelonephritis is made and it is decided that urgent investigation of his urinary tract is necessary. Pending the antibiotic susceptibility results on his blood and urine isolates, intravenous ciprofloxacin 400 mg bd is continued as most coliforms would be expected to be susceptible.

Infectious Disease: Clinical Cases Uncovered. By H. McKenzie, R. Laing, A. Mackenzie, P. Molyneaux and A. Bal. Published 2009 by Blackwell Publishing. ISBN 978-1-4051-6891-5.

What is acute pyelonephritis?

Acute pyelonephritis is an infection in the renal pelvis that may be associated with infection in the renal parenchyma. Acute pyelonephritis can result either from an ascending infection from the lower urinary tract or can be acquired by the haematogenous route. Gram-negative bacteria such as *Escherichia coli, Klebsiella* and *Proteus* are the common causes of acute pyelonephritis. Note that *E. coli* is by far the commonest cause of UTIs, but other common causes are listed in Table 21.1. Urinary catheters are often colonised by microorganisms, but colonisation does not require antibiotic treatment. Treat only if the patient is symptomatic or if catheter manipulation is necessary.

What is the significance of the 10^5/ml figure for urine culture?

Because it is difficult to distinguish between true infection and contamination of a urine culture, 10^5 organisms/ml has become a popular quantitative 'cut-off' for a significant culture. In fact, there is good evidence that lower numbers are significant in symptomatic infection, so any pure growth in the presence of pyuria (pus cells in urine) is likely to be significant in a patient with urinary symptoms. However, the 10^5/ml cut-off is useful when applied to asymptomatic patients, e.g. urine screening in pregnancy (see Case 15), when it does help to distinguish between contamination and infection.

Table 21.1 Common causes of urinary tract infections.

Organism	Comments
Escherichia coli	Commonest cause of UTI
Enterococcus faecalis	
Proteus species	Associated with stone formation
Klebsiella species	
Staphylococcus saprophyticus	Coagulase-negative staphylococci are not usually significant pathogens in UTI, but this species is an exception
Pseudomonas aeruginosa	Unusual in simple uncomplicated UTI. May be associated with previous antibiotic treatment or catheter-related infection
Candida albicans	Unusual in simple uncomplicated UTI. May be associated with previous antibiotic treatment or catheter-related infection

What other differential diagnoses should be considered in this patient?

Several intra-abdominal pathologies could mimic acute pyelonephritis. These include acute appendicitis, acute cholecystitis, an intra-abdominal abscess and, in female patients, ectopic pregnancy, endometritis and salpingitis. Acute pyelonephritis must be differentiated from a perinephric abscess. Abscesses can also form within the renal parenchyma. Abdominal imaging will help establish the diagnosis with certainty.

Why did the patient get a third UTI?

Recurrent attacks of UTI can be either a relapse or a re-infection. A relapse is a repeat infection with the same organism. A re-infection on the other hand is a new infection. Recurrent UTI often occurs in the presence of a congenital anomaly in the urinary tract or an obstruction to the normal flow of urine. Remember, a congenital abnormality may not manifest at birth. In fact, adult polycystic kidney disease, which is the commonest congenital abnormality of the urinary tract, manifests most commonly in the third or fourth decade of life. Obstruction of the urinary tract can be caused by a stone in the urinary tract or a stricture.

It is therefore important to look actively for predisposing factors in patients with recurrent UTIs. UTIs are in general much more common in females, but the prevalence increases in older males due to an increasing influence of prostatic hypertrophy.

What further steps would you take to look for predisposing factors?

Recurrent attacks of UTI should prompt a thorough investigation particularly in males. An ultrasound or a CT scan of the abdomen (Fig. 21.1) is indicated to find out the underlying pathology. CT scans of the abdomen are generally more sensitive than abdominal ultrasound.

> *A CT scan of the abdomen showed the presence of a 7mm stone in the ureter, along with a 6cm diameter fluid collection around the right kidney containing gas bubbles, circumscribed and well defined by a thickened fascia.*

What is a perinephric abscess?

The presence of an abscess in the renal parenchyma (renal abscess) or in the perinephric space (perinephric

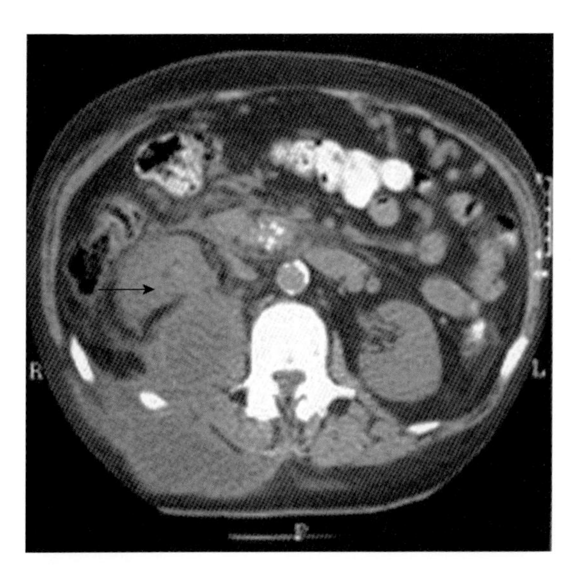

Figure 21.1 Example of a CT scan of the abdomen showing a perinephric abscess (arrow) around the right kidney. Courtesy of Professor F. J. Gilbert.

abscess) is a recognised complication of acute pyelonephritis. The perinephric space lies between the capsule of the kidney and the fascia of Gerota. This space contains fat and blood vessels. Thus, an abscess in this space is limited by the fascia. The abscess can also breach the fascia and develop into a paranephric abscess and cause peritonitis.

Perinephric abscess can be a complication of acute pyelonephritis or can develop following haematogenous spread from foci outside the urinary tract. An abscess that develops following haematogenous spread is commonly caused by *Staphylococcus aureus*. On the other hand, Gram-negative bacilli isolated from perinephric abscesses are more commonly due to local spread of infection from the renal calyces.

What should be done now?

Drainage is the usual way of treating a perinephric abscess and can be done either by a percutaneous nephrostomy or by open drainage. Some cases can be managed conservatively. Large abscesses should always be drained, particularly when associated with obstruction (with a urinary stone as in this case or by a stricture).

Urinary tract calculi can be treated either medically or surgically. Small stones usually pass through the renal tract on their own and hydration therapy can aid this process. Larger stones need to be removed surgically. Stones larger than 6 mm usually require surgical intervention. Smaller stones should also be removed surgically where medical therapy fails. Stones associated with infection in the proximal renal tract also require surgical removal. Stones can be of various types and may be composed of calcium oxalate, calcium phosphate, uric acid, struvite or cystine. Infection in the urinary tract, especially with *Proteus* species, is one of the precipitating factors for stone formation.

Various surgical procedures are available for removing ureteric calculi. These include extracorporeal shock wave lithotripsy and invasive procedures such as ureteroscopic removal with the help of a laser, percutaneous removal and even open surgery. A ureteral stent may be placed while one of these procedures is carried out in order to bypass the stone and aid the flow of urine into the urinary bladder.

The Gram-negative bacillus in the patient's blood and urine was identified as Proteus mirabilis sensitive to co-amoxiclav, cefotaxime, ciprofloxacin, piperacillin-tazobactam, meropenem and gentamicin. It was resistant to trimethoprim, nitrofurantoin and amoxicillin. The patient was continued on intravenous ciprofloxacin.

The surgical team was consulted and a decision was made to drain the perinephric abscess percutaneously by placing a nephrostomy tube in the right kidney. Five days later, the ureteric obstruction was relieved with the help of a J-stent that bypassed the ureteric stone, followed by removal of the nephrostomy tube. The stent provided a channel for the flow of urine from the renal calyces to the bladder. Daily bloods were done and urine output recorded to assess the renal function.

The patient improved following drainage and stenting. Five days following the procedure, intravenous ciprofloxacin was discontinued and the patient was started on oral ciprofloxacin 500 mg bd to finish a 14-day course. His fever subsided and his inflammatory markers began to improve.

How will you follow the case?

The emergency management has been successful but the underlying pathology must also be treated. At an appropriate time, the ureteric stone will have to be removed by one of the methods described above. Follow-up CT scans are indicated in order to assess the outcome. Complete regression of the lesions on a CT scan may take a few weeks.

The ureteric stone was removed with the help of extracorporeal shock wave lithotripsy 2 weeks following stenting of the ureter. It was found to be a struvite stone. Struvite stone formation occurs at a high pH and this explains the association with Proteus mirabilis, since this organism produces copious amount of urease which splits urea to produce ammonia. The patient was discharged and was followed up and when seen in the clinic after 4 weeks was well.

CASE REVIEW

A 53-year-old man presented with fever, back pain and abdominal pain. UTI was suspected and it transpired that he had had two previous admissions for UTI in the last 5 months. Blood cultures were positive for *Proteus mirabilis* in both bottles and a CT scan showed the presence of a ureteric stone and a perinephric abscess. The abscess was drained percutaneously and a stent inserted to bypass the stone. The infection was treated with ciprofloxacin and the stone later removed by lithotripsy. The patient made a complete recovery.

KEY POINTS

- Bacteraemia due to *E. coli* or other coliforms is often due to underlying infection in the urinary tract.
- Recurrent UTIs should be thoroughly investigated.
- Appropriate treatment of a complicated UTI often requires surgical input.
- *Proteus mirabilis* infection in the urinary tract is often associated with stone formation.
- Severe pain in a patient with an infection is an 'alert' symptom that implies tissue destruction and possible abscess development. Surgical intervention is probably required.

Further reading and information

SIGN Guideline 88. *Management of Suspected Bacterial Urinary Tract Infection in Adults*. Available at www.sign.ac.uk.

Case 22 A 20-year-old female student with a rash, fever, myalgia and diarrhoea

A 20-year-old student presented to the Accident and Emergency Department complaining of feeling unwell, headache, rigors, sweating and the development of a rash. She was concerned about a retained tampon which she had attempted to remove 2 days ago, but had only extracted half of it.

What differential diagnoses would you think of at this stage?

1 Toxic shock syndrome (TSS).
2 Measles.
3 Rubella.
4 Meningococcal disease.
5 Scarlet fever.

What else would you like to ask in the history?

Given the history of a retained tampon, there is a possibility of staphylococcal TSS. It is therefore important to enquire about symptoms such as diarrhoea, aching muscles, red eyes and confusion. However, a full systematic enquiry is necessary because the retained tampon may be a 'red herring' and this non-specific presentation may be due to a viral or a bacteraemic illness.

Past medical history

Any previous episodes of this type of illness, or the others listed above, would be important to know about.

Drug allergies

Has the patient taken any recent antibiotics which could have caused the rash?

Infectious Disease: Clinical Cases Uncovered. By H. McKenzie, R. Laing, A. Mackenzie, P. Molyneaux and A. Bal. Published 2009 by Blackwell Publishing. ISBN 978-1-4051-6891-5.

Family history

Do any close contacts have a similar illness?

On systematic enquiry, she does have myalgia, red eyes, dizziness, disorientation and diarrhoea. Her symptom complex would be consistent with staphylococcal TSS.

What are the key things to look for in a patient with TSS?

- Fever: a temperature of 39 °C or greater.
- Rash: diffuse, macular and erythematous. The rash particularly affects the peripheries, may contain petechiae, and may be exaggerated in the flexures. The rash typically desquamates after 1–2 weeks. Conjunctival injection is often prominent.

On examination she was dehydrated with a pulse of 96 beats/min, temperature 39.1 °C and BP 102/55 mmHg. There was a generalised, macular, erythematous rash (Plate 13; see Case 6 for a description of different types of rash). There was conjunctival suffusation. The chest, cardiovascular system and abdomen were normal. There was a thick, purulent discharge from the vagina, but no tampon or foreign body was present.

What is your working diagnosis now?

Probable staphylococcal TSS. The rashes of the other conditions listed in the differential diagnoses above are different:

- Measles and rubella: maculopapular rash.
- Meningococcal disease: petechial rash.
- Scarlet fever: fine, red, rough textured rash.

What investigations would you perform and why?

- Full blood count – thrombocytopenia is often present ($<100 \times 10^9$/ml).
- Urea and creatinine – creatinine is often twice the upper limit of normal.

- Liver function tests – elevated bilirubin and transaminases.
- C-reactive protein would be expected to be elevated.
- Vaginal swabs to culture the causative organism.
- Blood for detection of toxic shock syndrome toxin (TSST).
- Creatinine kinase, if elevated, suggests myositis.

Initial blood investigations

		(Normal range)
Haemoglobin	*12.9 g/dL*	*(11.5–16.5)*
White blood cell count	*29.8 × 10^9/L*	*(4–11)*
Platelet count	*94 × 10^9/L*	*(150–400)*
Serum urea	*9.3 mmol/L*	*(2.5–7.5)*
Serum creatinine	*116 μmol/L*	*(60–110)*
Serum alanine aminotransfrase	*112 U/L*	*(5–35)*
Serum alkaline phosphatase	*104 U/L*	*(45–105)*
Serum gamma glutamyl transferase	*102 U/L*	*(4–35)*
Serum creatinine kinase	*450 IU/L*	*(25–170)*
Serum C-reactive protein	*206 mg/L*	*(<10)*

What is the likely diagnosis at this stage?

The clinical presentation, leucocytosis, thrombocytopenia, renal dysfunction, hepatic dysfunction, myositis and elevated CRP all suggest a diagnosis of TSS. In this case the TSS is due to toxins produced by *Staphylococcus aureus*. TSS may also be caused, less commonly, by *Streptococcus pyogenes*.

TSS may be defined as a syndrome of fever, profuse macular erythematous rash with or without desquamation, a low systolic blood pressure (<90 mmHg) plus involvement of two or more of the following organ systems:

- Gut – vomiting and diarrhoea.
- Muscles – elevated creatinine kinase.
- Inflamed mucosae – vagina, mouth, conjunctiva.
- Renal dysfunction.
- Hepatic involvement.
- Blood – low platelets.
- Central nervous system – altered conscious level.

What is the pathology of TSS?

There is likely to be a staphylococcal infection, which may in itself appear trivial, but the systemic symptoms are toxin mediated. TSS has typically been associated with tampon use in healthy menstruating women. However, TSS has also been associated with non menstrual infections such as pneumonia, osteomyelitis, sinusitis and skin infections (scalded skin syndrome).

The exotoxin, TSST-1, is the major *Staphylococcus aureus* toxin responsible for the menstrual form of the disease, but many other staphylococcal toxins exist and are involved in other non menstrual forms. *Streptococcus pyogenes* can also cause TSS through the production of *S. pyogenes* exotoxins A and B. All of these toxins act as superantigens and stimulate the production of cytokines such as tumour necrosis factor, interleukins, M protein and interferon γ (see Infection and immunity, Part 1). Almost every organ system can then be involved, including the cardiovascular, renal, skin, mucous membranes, gastrointestinal, musculoskeletal, hepatic, haematological and central nervous systems. Blood cultures are typically sterile in TSS, the disease being mediated by toxin.

How should this patient be treated?

Based on the clinical suspicion, antibiotics should be commenced as soon as possible and given intravenously. The treatment choice is flucloxacillin, but a second drug such as rifampicin or fusidic acid may be added. If the patient has a true penicillin allergy, clindamycin, vancomycin, teicoplanin or linezolid would be suitable alternatives, the latter three also being appropriate choices if the illness was due to MRSA. It is therefore essential to take a vaginal swab in order to identify the organism and test for antibiotic sensitivity.

Are there any adjunctive therapies that may be used?

- Some patients with TSS are treated with intravenous immunoglobulin, which has been shown to be effective in neutralising the toxin and improving the outcome.
- Fluid resuscitation and oxygen therapy as for all sick patients.
- Depending upon the site of infection in TSS, surgical drainage of pus, if necessary including debridement, may be required.

What is the prognosis of TSS?

Staphylococcal TSS has a mortality of less than 3% whereas streptococcal TSS is more severe, with a mortality of 30–70%.

Progress

The vaginal swab confirmed a profuse growth of Staphylococcus aureus which was sensitive to flucloxacillin. A reference laboratory report later confirmed the presence of TSST-1 in the blood, confirming the diagnosis.

The patient responded well to a 1-week course of flucloxacillin and rifampicin and made a full and complete recovery. In the convalescent period she had marked desquamation of her digits.

CASE REVIEW

A 20-year-old female student presented to Accident and Emergency with headache, rigors and an erythematous rash. In view of a history of a retained tampon, diagnostic suspicion focussed on toxic shock syndrome (TSS), although other differential diagnoses were considered. As TSS is most often a toxin mediated staphylococcal disease, the patient was treated with flucloxacillin and rifampicin and re-hydrated. No tampon or foreign body was detected on vaginal examination but *Staphylococcus aureus* was cultured from a low vaginal swab and TSST-1 was later detected in blood by the reference laboratory. The patient made a full recovery.

KEY POINTS

- A febrile illness with a rash and a history of a retained tampon should alert you to the possibility of toxic shock syndrome.
- The diagnosis of TSS is a clinical one based on the presence of several clinical features.
- TSS is usually caused by a toxin produced by *Staphylococcus aureus* but can also be caused by *Streptococcus pyogenes*.
- TSS is treated with antibiotics. Some patients are also treated with intravenous immunoglobulin, others may require surgery.

Further information

The toxic shock syndrome information service provides information for both members of the public and the medical profession at www.toxicshock.com

Case 23 A 54-year-old man with a cough and night sweats

A 54-year-old barman was referred to hospital by his GP after developing a persistent cough and complaining of night sweats. In taking the history the following symptoms were elicited:

- *Cough: this had been present for 4 weeks and was productive of green sputum. He had received a course of oral amoxicillin during the second week of his illness but there had been no improvement.*
- *Haemoptysis: there had been several episodes when he had noticed some flecks of blood in his sputum.*
- *Weight loss: despite continuing on his usual diet, he had lost almost 5 kg during the month of his illness.*
- *Sweats: these were most troublesome at night and had prompted him to get up from bed and change his pyjamas and bedclothes on a number of occasions.*
- *Fatigue: he had been feeling increasingly tired during the course of his illness.*

What is there in the history that suggests to you that this may not be a typical chest infection?

The duration is rather longer than would be expected for a lower respiratory tract infection due to a 'typical' organism such as *Streptococcus pneumoniae* or *Haemophilus influenzae*. It is also a concern that the patient has coughed up blood on some occasions – this is a symptom that you must always investigate. Furthermore, the lack of response to a course of antibiotics suggests that there may be more to this than a simple chest infection.

What else do you want to establish from the history?

- Does the patient smoke?
- Has he had previous chest infections?
- Does he know of anyone with similar symptoms?

Infectious Disease: Clinical Cases Uncovered. By H. McKenzie, R. Laing, A. Mackenzie, P. Molyneaux and A. Bal. Published 2009 by Blackwell Publishing. ISBN 978-1-4051-6891-5.

He stopped smoking 10 years ago having smoked about 20 cigarettes per day for 30 years. He was exposed to passive smoke in the bar where he worked until the smoking ban was brought in a few years ago.

There is no previous history of respiratory illness and he cannot recall ever having had a chest X-ray.

The patient lives alone but is aware of one regular customer in the bar who has had a chronic cough which he puts down to asthma. He believes that the customer has refused to go and see his doctor about it.

There is not much else of note in the history except to say that the patient is a fairly heavy drinker, consuming around 40 units of alcohol per week.

Is there anything to find on examination?

The only positive findings were that he appeared pale and lean with some inspiratory crepitations audible bilaterally over the upper lungs. There was no finger clubbing and no lymphadenopathy.

Which investigations would you like to perform?

Any smoker with an unresolving respiratory infection needs to be investigated for the possibility of a bronchial carcinoma, so the first investigation to do is a chest X-ray (Fig. 23.1). The X-ray shows cavitation and fibrosis of the right upper zone. There are a number of conditions that can lead to cavitation including bronchogenic carcinoma, some forms of occupational lung disease and bacterial pneumonia caused by *Staphylococcus aureus* or *Klebsiella pneumoniae*. However, the commonest cause of this appearance is tuberculosis (TB) and this would be very much in keeping with the presenting history.

What needs to be done now?

TB is not always infectious but you should assume that it is until you can get some more information. The patient

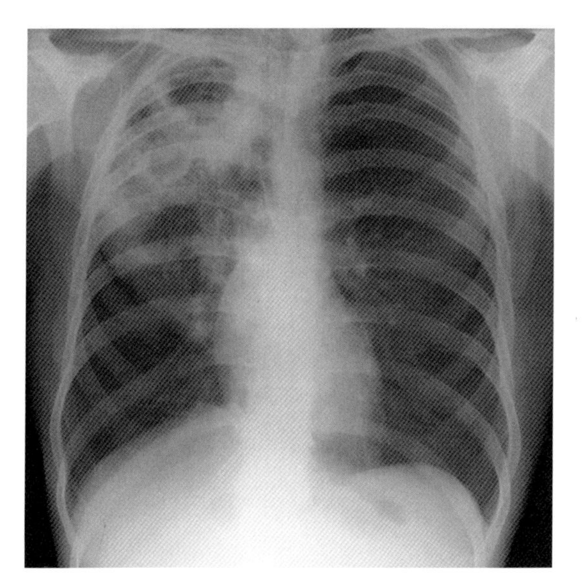

Figure 23.1 Chest X-ray showing cavitation and fibrosis involving the right upper and mid zones. Courtesy of Professor F. J. Gilbert.

Box 23.1 Adverse effects of the long-term use of anti-tuberculous drugs

Drug	Common adverse effect
Isoniazid	Hepatic toxicity Peripheral neuropathy – can be reduced by taking pyridoxine (vitamin B6) supplements
Pyrazinamide	Hepatic toxicity Hyperuricaemia – may lead to gout
Rifampicin	Orange-red discoloration of secretions, e.g. sweat, tears, urine Hepatic toxicity Drug interactions
Ethambutol	Optic neuritis – may present with red-green colour blindness or reduction in visual acuity

should be isolated in a single room and sputum samples obtained for auramine phenol or ZN stain to look for AAFB (see Laboratory diagnosis of infection, Part 1).

A sputum sample showed profuse AAFB on microscopy so it was appropriate to start antituberculous drugs without further delay. The patient was commenced on a combination of rifampicin, isoniazid, pyrazinamide and ethambutol. Her visual acuity and colour vision were checked before starting treatment because ethambutol can cause visual disturbance.

The logic behind using a combination of drugs is to prevent the emergence of resistant mutants which could appear quite easily if only one drug was used. The treatment for TB continues for at least 6 months so it is important to watch out for side effects that might occur with such prolonged therapy (Box 23.1).

Could the patient have HIV infection?

The development of tuberculosis may reflect immunodeficiency, so the patient should be counselled and tested for HIV infection. While most UK-born patients with TB do not have HIV infection, many TB patients born overseas do and this has major implications for their management. In HIV-infected patients, pulmonary TB may be more severe or unusual in its distribution and present as extrapulmonary TB (e.g. pleural, pericardial, peritoneal, meningeal, renal or spinal TB).

How do you know that the treatment is working?

Traditionally *Mycobacterium tuberculosis* has been cultured on solid media such as Lowenstein–Jensen medium and growth typically takes 4–6 weeks. However, laboratories increasingly use liquid media in automated systems (similar to blood culture bottles) that detect mycobacterial growth at an earlier stage of 2–3 weeks. It is only following growth in culture that it is possible to identify the organism as *M. tuberculosis* and find out whether it is sensitive to all or only some of the four drugs that the patient is taking. However, you would expect to get some clinical clues that your treatment is working. Within 2 weeks you would expect the fevers and sweats to have settled and for the patient's general well-being to have improved.

When might you be alerted to the possibility of drug-resistant TB?

- Failure to take medication properly increases the risk of developing resistant TB. In countries where drug compliance has become a problem, patients are observed taking their tablets (directly observed treatment or DOTS) to reduce the chance of resistance.
- Travel history: resistant TB has become a huge problem in the world so if the patient had spent time in countries where drug resistance is common (such as eastern Europe or South Africa), he would have been considered high risk for drug-resistant TB.

- If the patient is suspected of having multidrug resistant (MDR) or extensively drug resistant (XDR) TB, the patient should be nursed in a negative pressure room to reduce the risk of spread of infection to other patients or staff. Some strains are so resistant to treatment that there are very few drugs that are effective against them (Box 23.2).

Box 23.2 Definitions of durg-resistant tuberculosis

Term	Resistant to
Multidrug resistant (MDR) TB	Isoniazid and rifampicin
Extensively drug resistant (XDR) TB	Isoniazid and rifampicin, plus quinolones (e.g. ciprofloxacin, moxifloxacin, levofloxacin) and an injectable second-line agent (e.g. kanamycin, capreomycin, amikacin)

What else needs to be done?

Tuberculosis is a notifiable disease. The health protection team need to be informed of this case so that contacts can be screened for infection. This would include case finding amongst regulars in the pub, some of whom may require chest X-rays and review by the TB nurse specialist.

What is the outlook for the patient?

Sputum culture results came back in 6 weeks confirming that the M. tuberculosis strain was sensitive to all the standard treatment. The patient completed a total of 6 months' drug therapy – 2 months on all four drugs followed by 4 months on rifampicin and isoniazid. By the end of this treatment he was judged to have been cured of his tuberculosis and no further action was needed.

CASE REVIEW

A 54-year-old barman was referred to hospital with a 4-week history of persistent cough and night sweats. He also admitted to weight loss and haemoptysis. A course of amoxicillin prescribed by his GP had not made him feel any better. A chest X-ray was suggestive of tuberculosis and a sputum sample showed acid and alcohol fast bacilli on Ziehl–Neelsen stain. He was isolated in a single room and commenced on a combination of rifampicin, isoniazid, pyrazinamde and ethambutol. Six weeks later culture results confirmed infection with *Mycobacterium tuberculosis* which was fully sensitive to the above agents. The TB specialist nurse made arrangements to screen regulars at the pub where he worked and the patient made a full recovery after 6 months of treatment.

KEY POINTS

- An unresolving respiratory infection should be investigated for the possibility of a bronchial carcinoma and/or TB.
- Chest X-ray and a sputum sample for ZN stain (AAFB) and TB culture are the first steps. Bronchoscopy may be required to diagnose bronchogenic carcinoma.
- In any patient with persistent cough, night sweats and weight loss think about pulmonary TB. Do a chest X-ray and send a sputum for ZN stain and TB culture.
- Patients with AAFB-positive sputum should be placed in isolation and anti-tuberculous treatment commenced. Specialist referral is indicated as drug treatment can be complex, e.g. in multidrug-resistant TB.
- TB culture takes weeks and can be positive in AAFB-negative samples if only small numbers of organisms are present. *M. tuberculosis* cannot be fully identified or differentiated from other environmental or atypical mycobacteria on the basis of microscopy – it has to be cultured.
- Drug sensitivity tests are performed on cultures and results take some time to become available. However, some molecular tests that identify common resistance mechanisms are now available in reference centres.

Further reading and information

British Thoracic Society provides up-to-date information on tuberculosis at www.brit-thoracic.org.uk.

NICE Guidance. *Preventing and Treating Tuberculosis*. NICE, 2006. Available at http://www.nice.org.uk/nicemedia/pdf/CG033publicinfo.pdf.

Case 24 A 15-year-old boy with fever and a sore throat

A 15-year-old schoolboy attended his GP after developing a sore throat and fevers. He had been finding it increasingly uncomfortable to swallow and was struggling to eat or drink anything.

Which aspects of the history are most important?

Ask questions that might help to clarify the cause of the problem (this might include questions about previous tonsillitis or contact with anyone else who has had a similar illness). However, the immediate issue is the effect of this illness on the patient's upper airways. In relation to this you might want to ask:

- Is he having difficulty breathing?
- Has his breathing become noisy?
- Is he able to swallow anything – including his saliva?

He reports that his breathing is comfortable but he is unable to swallow anything, even saliva. He is able to speak but his voice is quite odd – almost as though he has something stuck in his throat.

What do you find on examination?

The oral cavity is best viewed with a bright torch and a tongue depressor.

On examination you see bilaterally enlarged tonsils coated with pus. You notice some petechiae on the hard palate. Although his breathing is noisy, there is no stridor. There is tender enlargement of the cervical lymph nodes.

Stridor is a high-pitched noise that is best heard when the patient breathes in – it is a sign of upper airways obstruction. If stridor had been present you would need to get help from an airways expert – an anaesthetist or ears, nose and throat surgeon – to make any necessary intervention in order to keep the airway open.

What tests would be helpful at this stage?

The main differential diagnosis is between a bacterial and a viral infection. A full blood count may prove very helpful in differentiating these. The results of a full blood count are shown below.

		(Normal range)
Haemoglobin	*13.7 g/dL*	*(13–18)*
White blood cell count	*12.5×10^9/L*	*(4–11)*
Neutrophil count	*5.9×10^9/L*	*(1.5–7)*
Lymphocyte count	*8.5×10^9/L*	*(1.5–4)*
Platelet count	*139×10^9/L*	*(150–400)*

The haematologist reports atypical lymphocytes on the blood film.

What does this all mean?

The blood test shows an elevated lymphocyte count and the film shows some cells are atypical in their appearance. This is therefore an illness associated with an atypical lymphocytosis. There are several infections associated with this condition:

- Glandular fever caused by acute Epstein–Barr virus (EBV) infection.
- Acute cytomegalovirus (CMV) infection.
- Acute toxoplasmosis.
- Acute HIV infection (seroconversion illness).

This presentation would be typical of glandular fever. The diagnosis can be confirmed by a heterophile antibody test (e.g. a Paul–Bunnell or monospot test) or by serology for EBV, which would be expected to show a

Infectious Disease: Clinical Cases Uncovered. By H. McKenzie, R. Laing, A. Mackenzie, P. Molyneaux and A. Bal. Published 2009 by Blackwell Publishing. ISBN 978-1-4051-6891-5.

positive IgM. The Paul–Bunnell test is based on the extraordinary finding that patients with glandular fever have antibodies against horse red cells. These are known as heterophile antibodies (*hetero*, other; *phile*, loving) and are directed against antigens that are common across a wide range of animal cells. This is a reflection of the deranged immune function that results from infection of B lymphocytes by EBV in the host, but in this case it provides a useful diagnostic test.

What do you need to do next?

Because the diagnosis is glandular fever, and therefore a viral illness, the patient does not need an antibiotic. The main treatment will be to keep the patient hydrated and relieve his symptoms with analgesia. Some patients with severe swelling of the tonsils can be given steroids to help reduce the swelling and speed up the rate at which breathing and swallowing improve. Not only are antibiotics unhelpful in glandular fever, some of them can cause harm. The semi-synthetic penicillins such as amoxicillin or ampicillin will cause a rash in most patients with glandular fever. This is not true penicillin allergy and does not recur once the patient has recovered from the infection. It is another example of the upset in immune function in glandular fever.

If the blood test had shown a high neutrophil count, you would have thought the illness was probably due to a bacterial infection. Now that diphtheria is unusual (with some exceptions worldwide), *Streptococcus pyogenes* is the only common bacterial cause of sore throat and the majority of sore throats are viral in origin. Intravenous benzylpenicillin or penicillin V (oral) would be appropriate choices of antibiotic if bacterial infection was suspected. Erythromycin or clarithromycin would be suitable alternatives if the patient was penicillin allergic.

Over the next 4 days the patient's swallowing improved and he was allowed home with advice to rest. He returned to school a week later and was asked to play for the first XV rugby team at the weekend. He managed the game with a bit of struggle but after a tackle he developed some abdominal pain that got progressively worse over the evening. His mother became concerned and took him to A&E.

On arrival at the hospital he appeared pale and uncomfortable. He was complaining of pain in the abdomen that was worse in the left upper quadrant with accompanying pain in the left shoulder. On examination he appeared pale and uncomfortable with a pulse of 115 beats/min, BP 90/45 mmHg and temperature 37.2 °C. His respiratory rate was 18 breaths/min and chest was clear to auscultation, but his abdomen was tender with guarding.

What needs to be done?

The picture is one of shock – he is hypotensive and tachycardic so urgent venous access should be obtained and fluid resuscitation given. Remember your ABC – i.e. check airway, breathing and circulation. The airway is not obviously compromised as he is talking. His respiratory rate is normal and chest clear to auscultation so there appear to be no breathing problems. His low BP and tachycardia are clearly indicative of a problem with his circulation, i.e. shock.

While you are resuscitating the patient you need to think about the cause. Shock is usually due to one of the following:

- Hypovolaemia, e.g. bleeding.
- Sepsis.
- Cardiogenic shock, e.g. following a myocardial infarction.

Cardiogenic shock is very unlikely in this setting and the history of sudden illness following trauma is not suggestive of sepsis, so hypovolaemic shock would seem most likely.

The A&E doctor called the on-call surgeon to assess the patient's abdomen and while awaiting his arrival the following blood count was received.

		(Normal range)
Haemoglobin	*7.7 g/dL*	*(13–18)*
White blood cell count	*13.5 × 10^9/L*	*(4–11)*
Neutrophil count	*10.9 × 10^9/L*	*(1.5–7)*
Lymphocyte count	*2.5 × 10^9/L*	*(1.5–4)*

The low haemoglobin supports the clinical suspicion that the patient may have intra-abdominal bleeding. The surgeon commenced a blood transfusion and arranged an urgent ultrasound scan, which showed features of a ruptured spleen. The patient was taken to theatre for an emergency splenectomy. He spent another week in hospital before being discharged home to convalesce.

Was the ruptured spleen linked to his recent glandular fever?

Glandular fever is associated with splenic enlargement, which can vary greatly in extent between patients. It is often possible to feel the tip of the spleen in a patient who presents with glandular fever and these patients should always be advised to avoid any contact sports for 6–8 weeks after their illness. This is because an enlarged spleen is more likely to rupture – sometimes spontaneously but more commonly following trauma. In this case, the tackle sustained during a rugby game would have caused his spleen to rupture and could have resulted in death if he had not undergone an emergency splenectomy.

Are there any consequences of having a splenectomy?

The spleen is the biggest lymphoid organ in the body, containing many of the body's immunoglobulin-producing B lymphocytes. When bacteria pass through the spleen they stimulate the production of opsonising antibodies, i.e. antibodies that attach to the pathogen and enhance phagocytosis. This function is particularly important in the clearance of encapsulated organisms (bacteria with a polysaccharide capsule) such as *Streptococcus pneumoniae, Haemophilus influenzae* type b and *Neisseria meningitidis.*

To try to reduce the chance of bacterial infection, patients undergoing splenectomy should be immunised against these three organisms. This is best done before splenectomy (if the operation is an elective one and not an emergency) since the antibody response to vaccination is better in patients with a spleen. In addition, patients can be given prophylactic antibiotics – usually oral penicillin V – following splenectomy to further reduce the risk of serious bacterial infection.

It is important to tell the patient that he should seek urgent medical attention if he ever develops a fever as untreated bacterial infection can prove rapidly fatal in splenectomised patients.

CASE REVIEW

A 15-year-old schoolboy presented with a sore throat and fever. He had been finding it increasingly uncomfortable to swallow and was struggling to eat or drink anything. Tests confirmed glandular fever caused by EBV infection. Despite being instructed to rest, he took part in a game of rugby a week or so later and was admitted to hospital shortly afterwards with a ruptured spleen. This required splenectomy and appropriate immunisation/follow up to minimise the increased risk of infection.

KEY POINTS

- Sore throats are most commonly viral.
- Glandular fever due to EBV is relatively common in young people and is associated with tiredness. It can be easily diagnosed by a blood film (atypical lymphocytes), by serology for EBV and by the monospot test.
- There are a few less common causes of the same clinical presentation, e.g. CMV, *Toxoplasma*, HIV.
- Amoxicillin should not be prescribed for a sore throat as patients with glandular fever nearly always develop a rash.
- Glandular fever is often accompanied by splenomegaly and there is an increased risk of traumatic rupture.
- Splenectomised patients should be immunised against capsulate organisms (*S. pneumoniae, N. meningitidis* and *H. influenzae*) and normally take prophylactic penicillin V.

Further reading and information

British Society for Haematology guidelines on the prevention and treatment of infections in patients with an absent or dysfunctional spleen are available at http://www.bcshguidelines.com/pdf/SPLEEN21.pdf.

Department of Health information for splenectomy patients is available at www.dh.gov.uk/en/Publicationsandstatistics/Publications/PublicationsPolicyAndGuidance/DH_4113581.

SIGN Guideline 34. *The Management of Sore Throat*. Available at http://www.sign.ac.uk/pdf/sign34.pdf.

MCQs

For each situation, please select the single best answer.

1 *An outbreak of diarrhoea and vomiting occurred in a nursing home over a 2-day period, affecting both residents and staff. It did not appear to be related to any particular meal and stool samples were sent for culture but no pathogenic bacteria were isolated.*

Which of the following is the most likely cause of the outbreak?

a. *Clostridium difficile*
b. Enterotoxigenic *Escherichia coli*
c. *Giardia lamblia*
d. Norovirus
e. Staphylococcal food poisoning

2 *A 75-year-old man was admitted to hospital with a 2-day history of abdominal pains and diarrhoea. He had been taking a course of amoxicillin for a respiratory tract infection for 1 week before the diarrhoea started. On arrival he had a fever of 38°C and his abdomen was distended and diffusely tender. An abdominal X-ray showed toxic dilatation.*

Which of the following treatments is likely to be of most value in treating the cause of his diarrhoea?

a. Cefotaxime
b. Oral ciprofloxacin
c. Gentamicin
d. Oral metronidazole
e. Prednisolone

Infectious Disease: Clinical Cases Uncovered. By H. McKenzie, R. Laing, A. Mackenzie, P. Molyneaux and A. Bal. Published 2009 by Blackwell Publishing. ISBN 978-1-4051-6891-5.

3 *A 52-year-old man was admitted to hospital with a 5-day history of fevers and left-sided pleuritic chest pain. He had no history of travel outside the UK in the previous year and had never before been a hospital patient. His chest X-ray showed an area of consolidation and Gram-positive diplococci were seen on microscopy in two blood culture bottles next day.*

Which of the following antibiotics would be most appropriate?

a. Benzylpenicillin
b. Gentamicin
c. Flucloxacillin
d. Metronidazole
e. Trimethoprim

4 *A 24-year-old injecting drug user was admitted to hospital with a 3-day history of bilateral pleuritic chest pain and fevers. An echocardiogram showed a vegetation on the tricuspid valve and his chest X-ray showed bilateral lung abscesses.*

Which of the following organisms is the most likely cause of the infection?

a. *Clostridium botulinum*
b. *Klebsiella pneumoniae*
c. *Staphylococcus aureus*
d. *Streptococcus pneumoniae*
e. *Streptococcus pyogenes*

5 A 60-year-old woman was admitted to hospital for treatment of a left lower lobe pneumococcal pneumonia. She received IV amoxicillin and showed clinical improvement with this – her fevers and pain resolved after 48 hours. One week after starting treatment she developed further fevers with her white cell count increasing from 8 to 25×10^9/L. She had only a peripheral IV cannula, which showed no signs of inflammation. A repeat chest X-ray showed a left-sided pleural effusion.

Which of the following is the most likely explanation for the patient's deterioration?

a. Empyema
b. Hospital-acquired pneumonia
c. Intravenous line infection
d. Pneumococcal resistance to amoxicillin
e. Pulmonary embolism

6 A 37-year-old man with untreated HIV infection was admitted with a 4-week history of a dry cough and increasing breathlessness. His chest X-ray showed bilateral lung infiltrates that mainly involved the perihilar areas.

Which of the following tests is most likely to reveal the diagnosis?

a. Blood cultures
b. Sputum PCR for pneumocystis
c. Lymph node biopsy
d. Mantoux test
e. Sputum culture

7 In a patient with symptomatic HIV infection, which of the following would be the most likely cause of retinitis?

a. *Candida albicans*
b. *Legionella pneumophila*
c. *Mycobacterium tuberculosis*
d. *Mycoplasma pneumoniae*
e. Cytomegalovirus

8 A blood culture from a patient with suspected endocarditis yields a staphylococcus from both bottles.

An initial classification of staphylococci can best be done by which of the following tests?

a. Haemolysis on blood agar
b. O serotype
c. Lancefield grouping
d. Coagulase test
e. Microscopic appearance

9 Which of the following statements about conjugate vaccines is true?

a. Conjugate vaccines stimulate passive immunity
b. Conjugate vaccines are given only once because they are live vaccines
c. Conjugate vaccines combine a carbohydrate antigen with a protein to improve the immune response
d. Conjugate vaccines contain toxoid
e. Conjugate vaccines should not be given to pregnant women

10 A 23-year-old man presented to his doctor with a 3-day history of worsening sore throat and pain on swallowing. Examination revealed bilaterally enlarged tonsils with exudates.

Which of the following bacteria is most likely to cause this illness?

a. *Haemophilus influenzae*
b. *Neisseria meningitidis*
c. *Staphylococcus aureus*
d. *Streptococcus pneumoniae*
e. *Streptococcus pyogenes*

11 A 15-year-old student was referred to hospital with a 24-hour history of fevers, headache and photophobia. On arrival she appeared drowsy with neck stiffness and a discrete purpuric rash over the trunk. A lumbar puncture was performed.

Which of the following types of organism are most likely to be seen on microscopy of the CSF after Gram staining?

a. Gram-negative bacilli
b. Gram-negative diplococci
c. Gram-positive cocci
d. Gram-positive bacilli
e. Mixed Gram-positive and Gram-negative cocci

12 *Infection with which of the following organisms would be most appropriately treated by metronidazole?*

a. *Bacillus anthracis*
b. *Clostridium difficile*
c. *Enterococcus faecalis*
d. *Haemophilus influenzae*
e. *Pseudomonas aeruginosa*

13 *Penicillins and cephalosporins act on bacteria by inhibition of which of the following?*

a. Cell wall synthesis
b. DNA synthesis
c. Protein synthesis
d. Flagellar function
e. All of the above

14 *Resistance to β-lactams in MRSA is due to which of the following?*

a. Altered penicillin-binding proteins
b. Increased β-lactamase activity
c. Increased efflux of antibiotic
d. Increased efflux of zinc
e. All of the above

15 *A 54-year-old man presented with fevers and a raised white cell count.*

Which of the following isolates is most likely to be clinically significant?
a. Coagulase-negative staphylococci in a wound swab
b. *Staphylococcus aureus* in a nose swab
c. *Escherichia coli* in a faeces sample
d. *Streptococcus pneumoniae* in a sputum sample
e. Viridans streptococci in a throat swab

16 *Phagocytosis is best described as which of the following?*

a. Activation of complement by IgG or IgM
b. Coating of the organism's surface by antibody and complement
c. Cytokine release following lymphocyte activation by lipopolysaccharides
d. Ingestion of organisms by polymorphs and macrophages
e. Intracellular killing of microorganisms

17 *Which of the following statements about capsulate organisms is true?*

a. Capsulate organisms are always Gram positive
b. Capsulate organisms are an important cause of overwhelming infection in splenectomised patients
c. Capsulate organisms are an important target for the cell-mediated immune response
d. Capsulate organisms are non-pathogenic to people with a normal functioning immune system
e. Capsulate organisms are resistant to intracellular killing

18 *Which of the following statements is correct in relation to microscopy of a Gram-stained specimen?*

a. If microscopy is negative, culture is not indicated
b. If microscopy is positive, culture will always be positive
c. Microscopy is considerably more sensitive than culture
d. Microscopy does not allow identification of the infecting species
e. Microscopy detects all bacteria present

19 *A 59-year-old man was referred to hospital with a 6-week history of weight loss, night sweats and a cough that was intermittently productive of blood-stained sputum.*

Which of the following investigations is most likely to yield the diagnosis?
a. Auramine phenol stain of sputum
b. Blood culture
c. Gram stain of sputum
d. Overnight culture of sputum
e. Serology for *Mycoplasma pneumoniae* infection

20 *A 32-year-old man was admitted to hospital with a 3-day history of fevers, headaches and myalgia. He had returned from a trip to West Africa 2 weeks before the onset of his illness.*

Which of the following pathogens is most likely to account for this presentation?

a. *Ascaris lumbricoides*
b. *Entamoeba histolyticum*
c. *Plasmodium falciparum*
d. *Schistosoma haematobium*
e. *Trypanosoma brucei*

21 *A 23-year-old woman presented to her doctor 2 weeks after returning from a gap year which she had spent in the Indian subcontinent. She had been vaccinated against hepatitis A 14 months earlier. She had no history of sexual contact or injected drug use and had been living in a camp in rural India. She compained of myalgia and had noticed herself to be jaundiced.*

Which of the following is the most likely cause of her illness?

a. Hepatitis A
b. Hepatitis B
c. Hepatitis C
d. Hepatitis D
e. Hepatitis E

22 *Benzylpenicillin is suitable empirical therapy for bacterial meningitis caused by which of the following organisms?*

a. *Escherichia coli*
b. *Haemophilus influenzae*
c. *Neisseria meningitidis*
d. *Pseudomonas aeruginosa*
e. *Staphylococcus aureus*

23 *Which of the following describes best the rationale for β-lactam and macrolide combination therapy for community-acquired pneumonia?*

a. They are synergistic against the common causes of pneumonia
b. The combination increases the spectrum of activity
c. Using combination treatment decreases the total duration of treatment
d. Use of combination allows lowering the dose of the individual agents thus limiting toxicity
e. The combination allows reduction in the frequency of administration

24 *A 24-year-old woman, who was 10 weeks' pregnant, presented 2 days after visiting her grandmother who was suffering from ophthalmic shingles. The woman asked if she was likely to get chickenpox as a result of this visit?*

Which of the following tests could be performed on the pregnant woman that would help to determine whether she was at risk of developing chickenpox?

a. Blood for varicella zoster virus PCR
b. Skin biopsy for viral culture
c. Throat swab for varicella zoster virus PCR
d. Varicella zoster virus IgG blood test
e. Varicella zoster virus IgM blood test

25 *A final year medical student set up a website offering travel advice to colleagues travelling to malarial areas to undertake their student electives. In his enthusiasm, he included some inaccurate information.*

Which of the following pieces of advice on malaria prevention is *not* appropriate?

a. Avoid travel to endemic regions if at risk of severe infection, e.g. splenectomised
b. Get vaccinated against malaria at least 4 weeks prior to the expected date of travel
c. Take daily doxycycline as anti-malarial prophylaxis
d. Use bed-nets impregnated with mosquito repellent
e. Wear loose fitting long-sleeved shirts

26 *Which of the following is consistent with bacterial vaginosis?*

a. Inflammation of the posterior vaginal wall on colposcopic examination
b. Presence of clue cells on microscopy
c. Presence of motile parasites (*Trichomonas vaginalis*) in the vaginal secretions
d. Presence of thick, white vaginal discharge
e. Vaginal discharge pH less than 4.0

27 *Which of the following infections is not sexually transmitted?*

a. *Chlamydia trachomatis*
b. Gonorrhoea
c. Hepatitis B
d. Rubella
e. Syphilis

28 *An 82-year-old man developed a fever 6 weeks after a total hip replacement. He had a history of chronic obstructive lung disease and had been on long-term steroid therapy. He had noticed worsening hip pain on the side of the prosthesis for 2 weeks. Following admission to hospital he had four sets of blood cultures that were positive for coagulase-negative staphylococci.*

What is the most likely explanation for the positive cultures?
a. Endocarditis
b. Lung abscess
c. Prosthetic hip joint infection
d. Nosocomial pneumonia
e. Skin contaminants

29 *A 44-year-old man developed a hoarse voice 5 days after starting a course of steroids for an exacerbation of his asthma. Examination of the oral cavity revealed white patches over the palate.*

Which of the following treatments would be most appropriate?
a. Aciclovir
b. Doxycycline
c. Fluconazole
d. Penicillin V
e. Rifampicin

30 *A 78-year-old woman was referred for assessment of fevers that had been ongoing for 6 weeks. She had also been aware of night sweats, left-sided headaches and fatigue. Following admission she was noted to have daily fevers of >38 °C. Her ESR was raised at 110 mm/h. A chest X-ray and CT scan of the chest/abdomen were normal. Four sets of blood cultures were negative.*

Which of the following is the most likely diagnosis?
a. Endocarditis
b. Lymphoma
c. Renal carcinoma
d. Tuberculosis
e. Temporal (giant cell) arteritis

EMQs

1 Infective causes of diarrhoea

a. *Campylobacter jejuni*
b. *Cryptosporidium parvum*
c. *Vibrio cholerae*
d. *Entamoeba histolytica*
e. Enterotoxigenic *Escherichia coli*
f. *Giardia lamblia*
g. Rotavirus
h. *Salmonella typhi*

For each of the following scenarios, choose the most likely causative organism from the options above. Each answer can be used once, more than once, or not at all.

1. A 24-year-old man developed diarrhoea and abdominal pains 48 hours after eating an undercooked chicken at a barbecue.
2. A 4-year-old boy was one of a group of ten children at a nursery to develop diarrhoea and vomiting shortly after the Christmas break.
3. A 19 year-old gap year student developed bloody diarrhoea while travelling in India. This settled after a week but 2 months later he developed right upper quadrant pain and fevers. An abdominal ultrasound scan revealed a liver abscess.

2 Infections of the mouth and nasopharynx

a. *Bacteroides melaninogenicus*
b. *Corynebacterium diphtheriae*
c. Coxsackie virus
d. Epstein–Barr virus
e. Herpes simplex virus
f. *Staphylococcus aureus*
g. *Streptococcus pyogenes*
h. Viridans streptococci

For each of the following scenarios, choose the most likely causative organism from the options above. Each answer can be used once, more than once, or not at all.

1. A 4-year-old child was admitted with fevers and sore throat which had started 3 days earlier. He had returned from eastern Europe the previous week and had no history of childhood immunisation. On examination he appeared pale, listless and unwell with a tachycardia of 140 beats/min. There was a grey adherent membrane over the tonsils and pharynx.
2. A 19-year-old student developed extensive painful ulceration over the lips and anterior mouth.
3. A 22-year-old woman with a history of recurrent tonsillitis was admitted with sore throat and fever. She had a left-sided peritonsillar swelling and tender left cervical lymph nodes.

Infectious Disease: Clinical Cases Uncovered. By H. McKenzie, R. Laing, A. Mackenzie, P. Molyneaux and A. Bal. Published 2009 by Blackwell Publishing. ISBN 978-1-4051-6891-5.

3 Antimicrobial chemotherapy

a. ... levels should be monitored during treatment
b. ... is used to treat HIV infection
c. ... is used to treat herpes simplex virus infection
d. ... acts only on Gram-negative bacteria
e. ... acts only on the yeast form of fungi
f. ... acts on protein synthesis
g. ... is a macrolide antibiotic
h. ... is a β-lactam antibiotic
i. ... is used in the treatment of bacterial vaginosis
j. ... is used in the treatment of athlete's foot

For each of the following, choose the most appropriate statement from the options above. Each answer can be used once, more than once, or not at all.

1. Zidovudine (a reverse transcriptase inhibitor) ...
2. Vancomycin ...
3. Ceftriaxone ...

4 Antimicrobial chemotherapy

a. Teicoplanin
b. Ceftriaxone
c. Aciclovir
d. Amoxicillin
e. Flucloxacillin
f. Nystatin
g. Nitrofurantoin
h. Amphotericin
i. Benzylpenicillin
j. Zidovudine

For each of the following, choose the most appropriate antimicrobial agent from the options above. Each answer can be used once, more than once, or not at all.

1. An agent active against herpes simplex virus
2. An antibiotic used in a combined preparation with clavulanic acid
3. An antifungal available for topical use only
4. A suitable agent for the treatment of MRSA infection
5. An agent used only for the treatment of urinary tract infection

5 Microbiological causes of infection and blood cultures

a. *Enterococcus faecalis*
b. *Escherichia coli*
c. *Haemophilus influenzae*
d. *Mycobacterium tuberculosis*
e. *Neisseria meningitidis*
f. *Salmonella typhimurium*
g. *Staphylococcus aureus*
h. *Staphylococcus epidermidis*
i. *Streptococcus pneumoniae*
j. *Streptococcus pyogenes*

For each of the following interim blood culture results, select the most likely causative organism from the list above. Each organism may be used once, more than once, or not at all.

1. A 66-year-old woman with diabetes and hypertension was admitted with a 5-day history of fever, cough and rigors. The chest X-ray showed consolidation in the right lower lobe. After 18 hours' culture, two blood culture bottles showed Gram-positive diplococci on microscopy.
2. A 46-year-old African man with HIV infection gave a 1-week history of fever, rigors and malaise. He reported a recent bout of diarrhoea but on presentation his abdomen was soft, although his right hip was tender and inflamed. His temperature was 39.5 °C. Twenty-four hours after admission, two blood culture bottles contained Gram-negative bacilli on microscopy.
3. A previously healthy 65-year-old man gives a 1-week history of fever, vomiting, jaundice, pale stool, dark urine and right upper quadrant pain. On examination there is tenderness in the right upper quadrant. After 24 hours of incubation, two blood culture bottles showed Gram-negative bacilli on Gram stain.
4. A 60-year-old man with diabetes gives a 3-month history of mid-thoracic back pain, fever, sweats and weight loss. An X-ray of his thoracic spine shows an abnormality in the T5 and T6 vertebral bodies. Several blood cultures were taken and, after 48 hours, Gram stain of four bottles showed Gram-positive cocci in clusters.

5. A 31-year-old woman is admitted with urinary frequency, dysuria and vomiting. On examination she was afebrile with normal pulse and blood pressure and no tenderness in the abdomen or loin. Forty-eight hours later the patient had been discharged home, but the laboratory phoned to say that one of two blood culture bottles submitted had became positive and showed Gram-positive cocci in clusters on Gram stain.

6 Antimicrobial prescribing

a. Amoxicillin
b. Benzylpenicillin
c. Ceftriaxone
d. Co-amoxiclav
e. Ciprofloxacin
f. Doxycycline
g. Metronidazole
h. Piperacillin-tazobactam
i. Trimethoprim
j. Vancomycin

For each of the following case scenarios statements, select the most appropriate antimicrobial therapy. Each item may be used once, more than once, or not at all.

1. A 21-year-old student presents with an 8-week history of offensive-smelling diarrhoea ever since a 3-week trekking holiday in Nepal.
2. A 19-year-old man presents with a 1-week history of severe sore throat and difficulty swallowing. On examination he has bilateral tonsillitis and a monospot blood test is negative.
3. A 36-year-old woman presents with a 1-week history of a generalised maculopapular rash. She also has a black eschar at the site of a tick bite which she received while on an overland trip in southern Africa.
4. A 70-year-old man with multiple recent hospital admissions, was admitted with fever, rigors and malaise and his blood cultures have grown *Staphylococcus aureus*. He has a documented severe allergy to penicillin and cephalosporins.
5. An 80-year-old woman with dementia, normally resident in a nursing home, was admitted to hospital with a 5-day history of profuse watery diarrhoea. She has received several courses of antibiotics for urinary tract infections in recent months. No other patients have been affected in her nursing home.

7 Investigations and laboratory techniques

a. Abdominal X-ray
b. Echocardiogram
c. Blood cultures
d. Lumbar puncture
e. Culture of bronchoalveolar lavage
f. Enzyme immunoassay for toxin in a stool specimen
g. Enzyme immunoassay for IgM in a serum sample
h. Nucleic acid amplification test of first void urine
i. Complement fixation test on acute and convalescent serum samples
j. Auramine phenol stain of sputum

For each of the following case scenarios, select the most appropriate investigation or laboratory technique from the list of options above. Each item may be used once, more than once, or not at all.

1. A 23-year-old female attending the genitourinary clinic for a *Chlamydia* screen.
2. A 75-year-old female with diarrhoea after 7 days of treatment with intravenous ceftriaxone.
3. A 45-year-old man in intensive care following abdominal surgery who has developed ventilator-associated pneumonia.

8 Clinical diagnoses

a. Typhoid
b. Malaria
c. Influenza A
d. Meningococcal meningitis
e. Osteomyelitis
f. Legionnaire's disease
g. Tuberculosis
h. Cellulitis
i. Hepatitis C
j. Septic shock

For each of the following case scenarios, select the most appropriate diagnosis from the list of options above. Each item may be used once, more than once, or not at all.

1. A 54-year-old man is admitted to the respiratory ward with pneumonia following a holiday in Spain. He presented initially with diarrhoea and shortness of breath. A urinary antigen test is positive.
2. A 45-year-old woman is admitted to intensive care with life-threatening *Staphylococcus aureus* pneumonia. Investigations reveal an underlying viral infection.
3. A 40-year-old oil worker was sent urgently to crisis manage a project in West Africa and had no time to go to the travel clinic for immunisations. On his return he was admitted with a severe flu-like illness, headache, constipation and fever. Blood cultures taken on admission revealed Gram-negative bacilli on Gram stain.

9 Infections associated with HIV disease

a. *Candida albicans*
b. Cytomegalovirus
c. *Cryptococcus neoformans*
d. *Cryptosporidium parvum*
e. Epstein–Barr virus
f. JC virus
g. *Listeria monocytogenes*
h. *Mycobacterium avium complex*
i. *Pneumocystis jirovecii (carinii)*
j. *Streptococcus pneumoniae*
k. *Toxoplasma gondii*

For each of the following clinical scenarios, choose the most likely causative organism from the options above. Each answer can be used once, more than once, or not at all.

1. A 23-year-old man with newly diagnosed HIV infection and a CD4 cell count of 18×10^6cells/L (normal range 430–1690) who presents with a 2-week history of blurred vision.
2. A 35-year-old man with untreated HIV infection who complains of pain on swallowing. Endoscopy reveals patchy white plaques in the oesophagus.
3. A 36-year-old woman with newly diagnosed HIV infection who presents with a 4-week history of a dry cough and increasing breathlessness.
4. A 55-year-old man with untreated HIV infection and a CD4 cell count of 56×10^6 cells/L (normal range 430–1690) presents with a 3-week history of frequent watery diarrhoea and cramping abdominal pains.

10 Prevention of infection

a. Passive immunisation
b. Source isolation
c. Antibiotic prophylaxis
d. Active immunisation
e. Live vaccine
f. Cohorting
g. Protective isolation
h. Epidemiological surveillance
i. Sterilisation
j. Disinfection

For each of the following scenarios, select the most appropriate option from those listed above that describes the action taken. Each answer can be used once, more than once, or not at all.

1. A new health worker starts a course of immunisation against hepatitis B.
2. A patient with suspected multidrug-resistant tuberculosis is barrier nursed in a negative pressure room.
3. A surgeon uses careful skin preparation with iodine in alcohol before inserting a central line.
4. Two patients with proven *Clostridium difficile*-associated diarrhoea were moved from the main ward to a two-bedded side room.

SAQs

1 A 17-year-old student is admitted to Accident and Emergency. Her flatmates were unable to rouse her this morning and called 999. Initial assessment shows that she is unresponsive, has a temperature of 38.5 °C and BP 90/50 mmHg. She has a purpuric rash on her trunk and neck stiffness.

a. What is the likely diagnosis and what would you do immediately? (*4 marks*)
b. If a lumbar puncture was performed, what would you expect to find in the CSF? (*4 marks*)
c. Who would you inform about this patient and why? (*2 marks*)

2 A 23-year-old woman attends your GP surgery complaining of a 2-day history of urinary frequency and dysuria.

a. What would you do? (*3 marks*)
b. What if she was pyrexial with loin pain and vomiting? (*3 marks*)
c. What if she had had three similar presentations in the last year? (*3 marks*)
d. What is the commonest cause of UTIs? (*1 mark*)

3 A 50-year-old man is admitted with shortness of breath and right-sided chest pain worse on inspiration. He is coughing up rusty coloured sputum. A chest X-ray shows consolidation of the right lower lobe.

a. How do you determine the clinical severity of pneumonia? (*2 marks*)
b. What specimens would you send to the microbiology laboratory to make a definitive diagnosis? (*4 marks*)
c. What antibiotic(s) might you prescribe if the patient had severe pneumonia and why? (*4 marks*)

4 A 30-year-old intravenous drug user is admitted with a fever and 1-week history of feeling generally unwell. On examination his chest is clear and he has no obvious soft tissue infection or injection abscesses. However, he has a murmur that is not noted in his previous medical records.

a. What is your suspected diagnosis? (*1 mark*)
b. List the clinical signs that might accompany this condition. (*3 marks*)
c. What investigations would confirm the diagnosis? (*2 marks*)
d. What is the commonest microbiological cause of this condition in drug users? (*1 mark*)
e. What is the commonest microbiological cause of this condition in non-drug users? (*1 mark*)
f. How would you treat this condition? (*2 marks*)

Infectious Disease: Clinical Cases Uncovered. By H. McKenzie, R. Laing, A. Mackenzie, P. Molyneaux and A. Bal. Published 2009 by Blackwell Publishing. ISBN 978-1-4051-6891-5.

5 A 60-year-old man is admitted with jaundice and right upper quadrant pain. He has a temperature of 39.3 °C and is tachycardic with a pulse of 130 beats/min and BP 75/40 mmHg which does not respond to IV fluids. He has an episode of rigors shortly after admission and urgent blood tests show an elevated white cell count, elevated urea and creatinine.

a. What are the criteria for septic shock and does this patient have it? (*3 marks*)
b. What urgent steps would you take? (*3 marks*)

The microbiology registrar telephones the following day to say that a blood culture taken at admission has Gram-positive cocci in both bottles, possibly an enterococcus.

c. List three antibiotics that would be suitable empirical therapy for an enterococcal infection. (*3 marks*)
d. The next day, the registrar calls again to say that tests show that the organism is a VRE. What does this mean? (*1 mark*)

6 A 70-year-old man attends your GP surgery for the third time in 3 months with a chest infection which has not responded to various antibiotics. He has lost 7 kg in weight over the last 6 months.

a. What is your differential diagnosis? (*2 marks*)
b. What investigations would you perform? (*4 marks*)
c. Assuming that the results are suggestive of pulmonary tuberculosis, what should happen now? (*4 marks*)

7 You are the microbiology registrar responsible for the blood culture bench this morning. Overnight, two sets of blood cultures (i.e. four bottles) from a 50-year-old male in the cardiology wards have all grown coagulase-negative staphylococci. The clinical details on the request form state only 'febrile – ?bacteraemic'.

a. What questions would you ask clinical staff when you telephone this result and why? (*5 marks*)
b. Assuming you confirm a genuine infection with coagulase-negative staphylococci, how would you treat it? (*5 marks*)

8 A 30-year-old drug user, known to be HIV positive, is admitted with jaundice. The duty doctor takes blood for hepatitis serology, but accidentally stabs herself with the needle before disposing of it.

a. The doctor is distressed and asks you for advice. What steps would you take? (*4 marks*)
b. What are the relative risks of transmitting hepatitis B, hepatitis C and HIV by needlestick injury from a patient known to be infected with that virus? (*2 marks*)

Results from the sample you sent from the patient are now available:

Hepatitis A IgM	*Positive*
Hepatitis B surface antigen	*Negative*
Hepatitis C antibody	*Positive*

In addition, you have now had the chance to review the patient's notes and found the following results dating from a previous admission 2 months ago:

Hepatitis C virus (HCV) PCR	*Negative*
HIV viral load	*<40 copies/ml*
CD4 count	*451 × 10^6 cells/L (normal range 430–1690)*
HIV resistance genotype	*No significant resistance*
HIV therapy	*Emtricitabine, tenofovir, efavirenz*

c. What does this mean for the patient and for the doctor?

9 A 45-year-old charity worker returns from a trip to Central Africa with a flu-like illness. He is fevered and increasingly unwell when admitted to the Infectious Diseases Unit for investigation. Initial examination confirms that he has tachycardia (110 beats/min), BP 104/68 mmHg and temperature 37.5 °C. However, there are no localising signs or symptoms.

a. What diagnoses would you like to exclude and how? (*4 marks*)
b. What preventative measures are available for these conditions? (*4 marks*)
c. List the likely causes of bloody diarrhoea in a returning traveller. (*2 marks*)

Case 10 boxes:

10 *A 54-year-old male undergoes a coronary artery bypass operation. Seven days postoperatively he develops a fever and tachycardia.*

a. What are the most likely sources of infection in a postoperative patient? (*3 marks*)

A sputum sample yields a profuse growth of MRSA and chest X-ray suggests bilateral lower zone pneumonia.

b. What are the options for treating MRSA pneumonia and what else would you do? (*3 marks*)

After 7 days of IV treatment, his chest is better, but he has developed significant diarrhoea and abdominal discomfort.

c. What test would you request? (*2 marks*)

d. This test is positive. What would you do now? (*2 marks*)

MCQ answers

1. d
2. d
3. a
4. c
5. a
6. b
7. e
8. d
9. c
10. e
11. b
12. b
13. a
14. a
15. d
16. d
17. b
18. d
19. a
20. c
21. e
22. c
23. b
24. d
25. b
26. b
27. d
28. c
29. c
30. e

Infectious Disease: Clinical Cases Uncovered. By H. McKenzie, R. Laing, A. Mackenzie, P. Molyneaux and A. Bal. Published 2009 by Blackwell Publishing. ISBN 978-1-4051-6891-5.

EMQ answers

1
1. a
2. g
3. d

2
1. b
2. e
3. g

3
1. b
2. a
3. h

4
1. c
2. d
3. f
4. a
5. g

5
1. i
2. f
3. b
4. g
5. h

6
1. g
2. b
3. f
4. j
5. g

7
1. h
2. f
3. e

8
1. f
2. c
3. a

9
1. b
2. a
3. i
4. d

10
1. d
2. b
3. j
4. f

Infectious Disease: Clinical Cases Uncovered. By H. McKenzie, R. Laing, A. Mackenzie, P. Molyneaux and A. Bal. Published 2009 by Blackwell Publishing. ISBN 978-1-4051-6891-5.

SAQ answers

1

a. Meningococcal meningitis seems highly likely. You would:
   - take a set of blood cultures but would not delay treatment unnecessarily
   - administer IV ceftriaxone
   - resuscitate with intravenous fluids and oxygen
   - get senior help; she is likely to need intubation, ventilation and inotropic support in the intensive care unit
   - not normally perform a lumbar puncture if the patient clearly has meningococcal septicaemia.

b. Gram-negative diplococci
   High white cell count, predominantly polymorphs
   CSF glucose would be reduced (less than 60% of blood levels)
   CSF protein would be grossly elevated.

c. The public health (health protection) department should be informed as they would consider antibiotic prophylaxis of close contacts.

See Case 5 for relevant background.

2

a. It is important from the history to ensure that this is a simple UTI (i.e. no loin pain, no systemic upset such as fever or rigors) and that there is no history of renal abnormalities, recurrent infections or underlying medical problems (e.g. diabetes). If you decide this is likely to be a simple infection, you might prescribe an antibiotic empirically, e.g. trimethoprim for 3 days. As a precaution, you might ask her to see the practice nurse to arrange for the submission of a mid-stream urine (MSU) – then if the organism turns out to be resistant to your choice of antibiotic, you can change it immediately.

b. Clinically, this sounds more like acute pyelonephritis. This requires 7–14 days of antibiotic therapy and you certainly want to check an MSU and make sure you are using an appropriate antibiotic. If the vomiting is significant, absorption of oral antibiotic could be impaired and she may need admission to hospital for 48 hours or so of intravenous antibiotics.

c. Recurrent UTIs should be investigated for the presence of underlying conditions that might predispose to infection. This should include checking her urine for glucose and imaging the renal tract to exclude an anatomical abnormality or the presence of stones.

d. *Escherichia coli.*

See Case 21 for relevant background.

3

a. The CURB-65 score.

b. Sputum for culture
   Blood culture
   Clotted blood for serological testing (follow up in 7–10 days)
   Urine for *Legionella* and pneumococcal antigen tests.

c. You would prescribe a β-lactam and a macrolide in combination. The β-lactam (e.g. co-amoxiclav or ceftriaxone) is there to cover *Streptococcus pneumoniae*, the commonest cause of lobar pneumonia. The macrolide (e.g. clarithromycin) is to cover the other so-called 'atypical' causes of pneumonia such as *Mycoplasma pneumoniae.*

See Case 7 for relevant background.

4

a. Infective endocarditis. Intravenous drug users characteristically present with infection of the

Infectious Disease: Clinical Cases Uncovered. By H. McKenzie, R. Laing, A. Mackenzie, P. Molyneaux and A. Bal. Published 2009 by Blackwell Publishing. ISBN 978-1-4051-6891-5.

tricuspid valve since the infecting organism is injected into the venous system.

b. The murmur of tricuspid regurgitation, or, less commonly, aortic or mitral regurgitation
   Splinter haemorrhages
   Embolic phenomena
   Roth spots.
c. Collect a series of blood cultures – say three sets over 24 hours, with extra careful skin preparation
   Echocardiogram.
d. *Staphylococcus aureus.*
e. Viridans streptococci.
f. In summary, treat with an appropriate bactericidal antibiotic(s) over an extended time period of weeks, not days. The correct answer is that once you had identified the infecting organism and obtained antibiotic susceptibility data, you would refer to published guidelines.

See Case 2 for relevant background.

5

a. Septic shock is defined as SIRS *plus* evidence of infection, organ dysfunction and hypotension despite adequate fluid resuscitation (as evidenced by a normal CVP using a central line). You do not have definitive evidence of infection yet or an assessment of his CVP, but septic shock seems likely. SIRS is recognised as the presence of two or more of the following:

Temperature	>38 °C or <36 °C
Heart rate	>90 beats/min
Respiratory rate	>20 breaths/min or $PaCO_2 < 4.3$ kPa
White blood cell count	$>12 \times 10^9$ cells/L or $<4 \times 10^9$ cells/L

b. He needs oxygen, a central line, intravenous fluids and antibiotics (take blood cultures first). You need senior help as he is likely to require transfer to the intensive care unit for possible ventilation, inotropic support and invasive monitoring.
c. Any three of the following: amoxicillin, piperacillin-tazobactam, vancomycin, teicoplanin or linezolid.
d. This is a vancomycin-resistant enterococcus (VRE). You need specialist microbiogy or infectious diseases advice on how to treat this as the antimicrobial options are limited.

See Antimicrobial chemotherapy (Part 1) and Case 17 for relevant background.

6

a. Bronchogenic carcinoma
   Tuberculosis
   Persisting lower respiratory tract infection.
b. Chest X-ray
   Sputum for Ziehl–Neelsen (or auramine phenol) and tuberculosis culture
   Consider CT scan of chest
   Consider bronchoscopy.
c. Source isolation of the patient in hospital, i.e. a single room, ideally negative pressure
   Commence anti-tuberculous therapy (with specialist advice)
   Inform the public health/health protection team.

See Case 23 for relevant background.

7

a. Coagulase-negative staphylococci are the commonest contaminants of blood cultures, but repeated isolation from more than one set of blood cultures must be taken seriously. Typically, coagulase-negative staphylococci do not cause infection in human tissue, but they cause infection of foreign bodies, so the question you must ask is: 'Does the patient have any long lines, prosthetic valves or a pacemaker *in situ*?'
b. Coagulase-negative staphylococci are usually resistant to flucloxacillin and therefore you would normally have to use a glycopeptide, e.g. vancomycin or teicoplanin. The problem is that it is very difficult to clear infection from the surface of the foreign body and it may have to be removed or replaced.

See Case 19 for relevant background.

8

a. Wash the wound in running water and encourage bleeding. Ask the patient for permission to have his blood sample tested for HIV and hepatitis C viral load in addition to the diagnostic tests you had planned for diagnosis of his jaundice (hepatitis A, B and C). Refer the doctor urgently to the infectious diseases department for consideration of immediate HIV post-exposure prophylaxis. From the patient's notes, some information might already be available (e.g. hepatitis B and C virus serology, CD4 count, HIV viral load, HIV resistance profile and current therapy). These results will help in deciding the best plan for the doctor. Ask the doctor to attend the occupational health department for follow up.

b. Hepatitis B 30%
   Hepatitis C 3%
   HIV 0.3%.

c. The patient's jaundice is almost certainly due to acute hepatitis A infection, which requires only symptomatic treatment. This result has no particular significance for the doctor. Although the hepatitis C antibody is positive, the HCV PCR is negative, which implies that the risk of infection is negligible as there is no active HCV replication. The patient's HIV infection is well controlled which reduces the risk of transmission. Nonetheless, the doctor is likely to be offered a 1-month course of anti-retrovirals as post-exposure prophylaxis against HIV. She will require specialist follow up for up to 6 months to ensure she has not been infected with either hepatitis C or HIV.

See Case 4 for relevant background.

9

a.
- Malaria: send blood to haematology for malaria antigen testing and/or for thick and thin blood films
- Typhoid: take blood cultures, preferably two or three sets over several hours. This is the best way to make the diagnosis, but stool and urine culture are also worthwhile.

b.
- Malaria: appropriate chemoprophylaxis, bed-nets and use of mosquito repellent
- Typhoid: immunisation, careful selection of food and bottled water only.

c. Causative organisms that are common in the UK can also be found abroad, so *Campylobacter* and *Escherichia coli* O157 are still possible. However, you would also be considering the possibility of *Shigella* species (e.g. *S. flexneri* or *S. dysenteriae*) and amoebic dysentery (*Entamoeba histolytica*).

See Cases 1 and 10 for relevant background.

10

a. Wound infection – in this patient, check the sternotomy and leg vein wound sites
   Pneumonia
   Urinary tract infection.

b. A glycopepetide (vancomycin or teicoplanin) or linezolid for treatment. Barrier nurse in a single room.

c. Send a stool sample for *Clostridium difficile* toxin (enzyme immunoassay).

d. Stop the IV treatment for his chest infection if it is no longer required. Start oral metronidazole.

See Cases 3 and 14 for relevant background.

Index of cases by diagnosis or organism

Case 1 *Escherichia coli* O157:H7, 34
Case 2 Infective endocarditis, 38
Case 3 Hospital-acquired MRSA pneumonia, 42
Case 4 Hepatitis A, 45
Case 5 Pneumococcal meningitis, 49
Case 6 Varicella zoster virus, 57
Case 7 Community-acquired pneumococcal pneumonia, 61
Case 8 Parvovirus B19, 66
Case 9 Influenza, 70
Case 10 *Plasmodium falciparum* malaria, 74
Case 11 Fever of unknown origin – adult Still's disease, 78
Case 12 Cellulitis, 82
Case 13 Genital infection with *Candida albicans* and *Chlamydia trachomatis*, 86
Case 14 *Clostridium difficile* and norovirus infection in a ward outbreak, 89
Case 15 Antenatal screening – positive hepatitis serology, 92
Case 16 HIV infection complicated by cytomegalovirus retinitis and cerebral toxoplasmosis, 96
Case 17 Gram-negative bacteraemia complicating ascending cholangitis, 101
Case 18 Leptospirosis, 106
Case 19 Prosthetic joint infection with coagulase-negative staphylococci, 109
Case 20 Febrile neutropaenia caused by pulmonary aspergillosis, 112
Case 21 Acute pyelonephritis complicated by a perinephric abscess, 115
Case 22 Staphylococcal toxic shock syndrome, 119
Case 23 Tuberculosis, 122
Case 24 Glandular fever due to Epstein–Barr virus complicated by a traumatic splenic rupture, 125

Index

Note: page numbers in *italics* refer to figures, those in **bold** refer to tables and boxes.

abdominal bleeding 126
abdominal pain
  diarrhoea 128
  fever 115–18
abscess
  lung 61, 62, 128
  perinephric 116–18
aciclovir 19
  shingles treatment 58
  varicella zoster treatment 58, 59
acid and alcohol fast bacilli (AAFB) 2, 123, 124, Plate 2
acquired immunity 23–4
acute myelogenous leukemia (AML) 112–14
adefovir 20
adolescents, rash 66–9
adult respiratory distress syndrome (ARDS) 52
Aeromonas hydrophila 84
AIDS *see* HIV infection
alcohol use/abuse
  acute pancreatitis 42
  hepatitis 45, 46
  HIV risk 75
allergy 31
  penicillin 84
  rash 66, 67
allylamines 21
aminoglycosides **16,** 17
  streptococcal resistance 14
amoebiasis 6
amoebic dysentery
  gastroenteritis 35
  hot stool examination 36
  see also Entamoeba histolytica
amoxicillin 15, **16**
amoxicillin–clavulanic acid *see* co-amoxiclav
amphotericin B 20
anaemia 102
antenatal screening 92–5
anthrax **33**
antibiotic(s) 13–15, **16,** 17–19
  allergies 31
  ascending cholangitis 104
  bactericidal 13
  bacteriostatic 13
  breakpoints 13
  cellulitis treatment 84
  fever treatment 80
  groups 15, **16,** 17–19
  infected hip prosthesis 110
  long-term prophylaxis 84, 85
  meningitis 52, 54, 55, 131
  meningococcal sepsis 51
  minimum bactericidal concentration 13
  minimum inhibitory concentration 13, *14*
  neutropaenia 113
  OPAT 110
  pneumonia 63, 64, 65
  prophylactic after splenectomy 127
  resistance 13–14
    mechanisms 14, **15,** 18
  S. aureus infection 43
  susceptibility 13–14
  toxic shock syndrome 120, 121
  see also named drugs and drug groups
antibiotic susceptibility tests 3–4
antibodies
  heterophile 126
  opsonising 127
antifungal drugs 20–1, 113
antigen(s), carbohydrate 27
antigen detection 7
  bacterial 4
  viral 5
  viral respiratory infections 71
anti-herpes virus drugs 19
anti-HIV drugs 19–20, 99, 100
antimicrobial chemotherapy 13–21, 134
  antibiotics 13–15, **16,** 17–19
  antiviral drugs 19–20
  prescribing 135
  see also named drugs and drug groups
antituberculous drugs 123, 124
antiviral drugs 19–20
  HIV infection 19–20, 99, 100
  resistance 20
approach to patient 28, **29–30,** 30–3
  diagnosis 32
  examination 32
  finishing consultation 32–3
  notifiable diseases 33
Aspergillus (aspergillosis)
  antifungal therapy 21
  diagnosis 114
  pulmonary 113
  serum antigen test 114
auramine phenol 2
autoimmune conditions, hepatitis 45, 46
azithromycin 18, 87
azoles 20–1
AZT (zidovudine) 19, 94

B lymphocytes 23–4
back pain, severe 115–18
bacteraemia
  fever 101, 102
  meningitis 53–4
  source 39
bacterial endocarditis 38, 39–41, 51
  aetiology 40
  diagnosis 39–40
  Duke criteria 40
  native mitral valve 40–1
  staphylococcal 129
  stigmata 102
  treatment 40–1
bacterial infections 1–5, **12**
  antigen detection 4
  culture 2–3, **4,** Plate 3
  diagnostic approaches **12**
  microscopy 1, 2
  nucleic acid detection 4–5
  serological testing 5
  specimen journey 1–5
  toxin detection 4
  see also antibiotic(s); *named organisms*
bacterial vaginosis 86, 132
barrier nursing, MRSA 28, 43
benzylpenicillin 15, **16**
  bacterial endocarditis 41
  cellulitis treatment 84
  leptospirosis 108
  meningitis treatment 52, 54, 55, 131
  pneumonia 64, 65
beta-lactam(s) 15, **16,** 17
  community-acquired pneumonia 131
  resistance 130
beta-lactamase 43
  see also extended spectrum beta-lactamase (ESBL)
biliary sepsis 103–5
blood culture 134, 138
  contaminants 39
  meningitis 52
blood films
  malaria 76, Plate 10
  parasitic infections 6
blood transfusion, abdominal bleeding 126
bone marrow, chemotherapy effects 112

Borrelia burgdorferi
  cellulitis differential diagnosis 82, 83
  laboratory diagnosis 5
breastfeeding
  hepatitis B virus 92
  HIV infection 94
bronchitis, influenza complication 71
bronchoalveolar lavage (BAL) 42
bronchoscopy, fibreoptic 114
Brucella (brucellosis) 107

calculi, urinary tract 117–18
Campylobacter, stool culture 36
Candida albicans (candidiasis)
  antifungal therapy 21
  HIV-associated 96, 97, **98,** 99, 100, Plate 12
  laboratory detection 6
  oral 96, 97, 132, Plate 12
  vaginal thrush 86, 87–8
capsulate organisms 130
carbapenems **16,** 17
carbohydrate antigens 27
cardiogenic shock 25
caspofungin 21
$CD4^+$ T cells 24
  count for HIV status 98, 99, 100
$CD8^+$ cells 24
ceftriaxone **16,** 17
  cellulitis treatment 84
  meningitis treatment 52, 54, 55
cefuroxime 17
cell-mediated immunity 24
cellulitis 82–5, Plate11
  bacterial causes 83–4
  examination 83
  investigations **84**
  necrotising 85
  prevention 84
  progress 84
  skin infection differential diagnosis 85
  treatment 84
central nervous system (CNS) examination 51
cephalosporins **16,** 17, 130
cerebrospinal fluid (CSF)
  examination in HIV infection 98
  meningitis lumbar puncture 52–3
chemoprophylaxis, travel history 74
chemotherapy
  antimicrobial 13–21, 134
    prescribing 135
  follow-up of patients 113
  infection risk 112
  remission–induction 112, 114
chest infection
  chest wall skin/soft tissue infection 61, 62
  weight loss 138
chest pain 137
  pleuritic 61–5, 128
chest X-ray
  malaria 76
  pneumonia 63, 65
  respiratory infection in smoker 122, *123*
chickenpox 57–60
  pregnancy 131
  rash 59, Plate 9
  susceptibility of contacts 59
children
  hepatitis B virus screening 48
  respiratory tract viral infections 73
Chlamydia psittacci 5
Chlamydia trachomatis
  antenatal testing **95**
  cervical infection 86–8
  complications 87
  laboratory diagnosis 4
  screening 87
cholangitis, ascending 103–5
  treatment 104–5
cholera **33**
chromosomal exchange/mutation 14
chronic obstructive pulmonary disease (COPD) 38
cidofovir 19
ciprofloxacin 18
  acute pyelonephritis 115–18
clarithromycin **16,** 17–18
  plus co-amoxiclav 63, 65, 106
clindamycin **16,** 18, 77
  cellulitis treatment 84
  Clostridium difficile-associated diarrhoea 90
Clostridium difficile
  enzyme immunoassay 36
  infection control measures 44
  laboratory diagnosis 4
  toxins 89
Clostridium difficile-associated diarrhoea (CDAD) 35, 44, 84
  027 ribotype highly virulent strain 90
  clindamycin 90
  enzyme immunoassay 89
  isolation 90
  outbreak 89–91
clotrimazole, candidiasis treatment 86, 87, 88
coagulase test 2–3, Plate 4
coagulation cascade, endotoxic shock 25–6
co-amoxiclav 15, **16,** 17
  plus clarithromycin 63, 65, 106
coliform bacteria 3
colitis, ischaemic/pseudomembranous 35
colonic carcinoma 35
complement 24
  activation in endotoxic shock 25–6
complement cascade 24
complement fixation test (CFT) 10–11
compression stockings 84
computed tomography (CT)
  abdominal for urinary tract infection 116, *117*, 118
  chest for pneumonia 113, 114
  hip prosthesis infection 109, 111
  HIV infection 97
consciousness, reduced 50–1
contacts
  tuberculosis case finding 124
  varicella zoster virus 58
coronary artery bypass, postoperative fever 139
costochondritis, fever with pleuritic pain 61, 62
cough 39
  chronic obstructive pulmonary disease 38
  fever and myalgia 106–8
  night sweats 122–4
  weight loss and night sweats 130
Coxiella burnetti 107
  laboratory diagnosis 5
C-reactive protein 24, **25**
Cryptosporidium parvum 6
culture
  bacterial species 2–3, Plate 3
  viruses 5
  see also blood culture
CURB-65 score 64
cytarabine 112
cytokines 24
  pro-inflammatory 25
cytomegalovirus (CMV)
  fever and sore throat 125
  hepatitis 45, 46
  laboratory detection 5
  retinitis 98
    treatment 99, 100
cytopathic effect (CPE) 5

daptomycin 19
  cellulitis treatment 84
deep vein thrombosis 82, 83
dehydration
  diarrhoea-induced 35, 36
  hypotension 51
delivery
  hepatitis B virus 92
  HIV infection 94
dengue fever **31**
diabetic ketoacidosis, influenza complication 71
diagnosis, clinical 136
diarrhoea
  abdominal pain 128
  antibiotic-associated 44
  bloody 34–7
  causes 133
  dehydration 35, 36
  differential diagnosis 34–5
  examination 35
  history taking 34
  imaging 36
  MRSA pneumonia 139
  rehydration 35, 89
  stool cultures 36
  viral 36
  and vomiting
    nursing home 128
    orthopaedic ward 89–91
  see also Clostridium difficile-associated diarrhoea (CDAD)
directly observed treatment (DOTS) 123
disc diffusion test 13–14
disseminated intravascular coagulation (DIC) 102
  purpuric rash 50

diverticulitis 35
doxorubicin 112
doxycycline 18, 77
drug reaction, rash 66, 67
drug use/abuse
  heart murmur 137
  hepatitis 45, 46
  HIV risk 75
  jaundice 138
  pleuritic chest pain with fever 128
dysuria 137

ear infection 51
echinocandins 21
ectopic pregnancy 87
elderly people, fever 38–41
encephalitis 32
endocarditis 51
  stigmata 102
  see also bacterial endocarditis
endoscopic retrograde cholangiopancreatography (ERCP) 104, 105
endotoxic shock 25–6
endotoxins 22, 25
  complement cascade 24
Entamoeba histolytica (amoebic dysentery)
  hot stool examination 36
  laboratory diagnosis 6
  travel history **31**
  see also amoebic dysentery
entecavir 20
Enterobacteriaceae 3
  antibiotic resistance **15**
enterococci 138
  antibiotic resistance **15,** 44
enterohaemorrhagic *E. coli* (EHEC) 36–7
enteroviruses
  hand, foot and mouth disease 60
  rash 57
enzyme immunoassays 4, 7, 8, 11–12
  Clostridium difficile 89
  HIV testing 97
epitopes 23, 24
Epstein–Barr virus (EBV)
  glandular fever 125–7
  hepatitis 45, 46
  laboratory detection 5
erythema
  infectiosum 67
  nodosum 82, 83
erythromycin 17–18
Escherichia coli 3
  acute pyelonephritis 116
Escherichia coli O157:H7
  diagnosis 36–7, Plate 8
  gastroenteritis 34, 35
  stool culture 36
ethambutol 123, 124
exotoxins 22
exposure prone procedures (EPPs) 48
extended spectrum beta-lactamase (ESBL) **15,** 44
extracorporeal shock wave lithotripsy (ESWL) 117–18

famciclovir 19, 102
fetus, varicella zoster virus complications 58
fever 51
  abdominal examination 102
  acute myelogenous leukemia 112–14
  antibiotic therapy 80
  with back and abdominal pain 115–18
  with cough and myalgia 106–8
  diagnosis 76
  differential diagnosis 107
  duration 78
  elderly man 38–41
  examination 38, 75, 78, **79,** 101–2
  headache
    and myalgia 131
    with night sweats 132
    and photophobia 129
  history taking 38, 78, **79,** 106
  investigations 38–9, 102–3
  muscle pains 70–3
  myalgia 106–8, 131
  pleuritic pain 61–5, 128
  post-coronary artery bypass 139
  raised white cell count 130
  recurrent 78–80
  with rigors, sweating and malaise 101–5
  with sore throat 125–7
  toxic shock syndrome 119
  travel/travel history 74–7, 138
fever of unknown origin (FUO) 78–80
  investigations 79
fifth disease 67
flucloxacillin **16,** 17
  toxic shock syndrome 120, 121
fluconazole 20–1, 86
fluid resuscitation
  pneumonia 62–3
  toxic shock syndrome 120
  see also intravenous (iv) fluids
fluoroquinolones 18
focal neurological signs 51
foscarnet 19
fungal infections 6
  cell-mediated immunity 24
  drug therapy 20–1
  specimen journey 6
  see also named organisms

gallstones 103–5
ganciclovir 19
Gardnerella vaginalis 86
gastroenteritis, bloody diarrhoea 34–5
gentamicin **16,** 17
  bacterial endocarditis 41
  monitoring use 41
  neutropaenia 113, 114
Giardia lamblia (giardiasis)
  laboratory diagnosis 6
  travel history **31**
glandular fever 125–7
  splenic enlargement 126–7
Glasgow Coma Scale 50–1
glycopeptide intermediate *Staphylococcus aureus* (GISA) **15**
glycopeptides **16,** 17

gonorrhoea 86, **95**
Gram stain 2, 130, Plate 1
Gram-negative pathogens 3, Plate 5
  acute pyelonephritis 115–18
  perinephric abscess 117
  vancomycin resistance 14
Gram-negative shock 24
Gram-positive pathogens
  classification 2–3
  diplococcus 53, 54, 64
  enterococcal 138
  fever 39
Gram-positive shock 26

haemolysis of blood agar 3, Plate 5
haemolytic uraemic syndrome 36–7, Plate 8
  treatment 37
Haemophilus influenzae
  immunisation after splenectomy 127
  laboratory diagnosis 4
  opsonisation 23
Haemophilus influenzae type b vaccine (Hib) 27, 55
  after splenectomy 127
haemoptysis, cough and night sweats 122
haemorrhoids, bleeding 35
hand, foot and mouth disease 59–60
headache 96–100
  differential diagnosis in HIV disease 97
  fever
    and myalgia 131
    and night sweats 132
    and photophobia 129
  severe with drowsiness 49–55
    history taking 49–50
    with nausea, vomiting and photophobia 50
Health Protection Agency 33
  measles notification 67
  meningitis notification 55
  rubella notification 67
healthcare workers
  blood-borne viral infections 48
  chickenpox contact 58, 59
  influenza pandemic 72
heart murmur 137
  systolic 38, 39
hepatitis
  causes 45
  contacts 47
  history 45–6
  immunisations 48
  testing 46, **47**
  travel history 131
  vaccines 48
  virology investigations 46
hepatitis A virus 45, 46, 48
  antenatal testing 92
  diagnosis 46–7
  IgM investigation 46, **47**
  jaundice in drug user 138
  travel history **31**
hepatitis B immunoglobulin (HBIG) 48
  babies of e-antigen positive mothers 92, 95

hepatitis B surface antigen (HBsAg) 27, 46, **47**
  antenatal testing 92
hepatitis B virus 45, 46–8
  antenatal testing 92, 93
  diagnosis 46–7
  drug therapy 20
  mother-to-child transmission 93
  pregnancy 47, 48, 92, 95
  referral 47
  screening of children 48
  testing 46, **47**
hepatitis C virus
  antenatal testing 92, 93, **95**
  antibody testing 46
  drug therapy 20
  HIV co-infection 94
  jaundice in drug user 138
  laboratory detection 5
  liver disease assessment 93, 95
  mother-to-infant transmission 93
  pregnancy 93, 95
  RNA testing **47**
hepatitis E virus 45, 46
hepatitis normal immunoglobulin (HNIG) 47, 48, 67
herpes simplex virus (HSV)
  antenatal testing **95**
  rash 57
herpes viruses
  antiviral drugs 19
  Kaposi's sarcoma 99
heterophile antibodies 126
hip prosthesis/joint pain, infected 109–11, 132
history taking 28, **29–30,** 30–2
  allergies 31
  isolation issues 28, **29**
  medication use 31
  presenting complaint 28, 30
  social history 31
  travel history 31
HIV infection 96–100
  antenatal screening 93–4
  antiviral drug resistance 20
  breathlessness 129
  cerebral toxoplasmosis 97, 100
  cough 129
  drug therapy 19–20
  fever and sore throat 125
  hepatitis C co-infection 94
  jaundice in drug user 138
  laboratory assessment of status 98
  laboratory detection 5
  mother-to-child transmission 94
  opportunistic infections 98, 136
  pregnancy 93–4
  prophylaxis 20
  retinitis 96, 97, 98, 129
  risk in returning traveller 75
  testing 46, 97
  tuberculosis 123
host immunity 22–4
human herpes virus 8 (HHV-8) 99
humoral immunity 23–4
hypotension 51
  management 54
hypovolaemic shock 25, 126

immunisation 26–7
  active 26–7
  hepatitis 48
  influenza 72
  passive 26
  pre-travel 74
  varicella zoster 59
  see also vaccines
immunity 22–4
immunocompromised patients
  parvovirus B19 68
  varicella zoster virus complications 58
immunofluorescence assay 7, 11–12
  respiratory syncytial virus 5, Plate 7
  viral respiratory infections 71
immunoglobulin(s) 23
immunoglobulin G (IgG) 23
  serological testing 10
immunoglobulin M (IgM) 23
  serological testing 10
infection(s)
  approach to patient 28, **29–30,** 30–3
  emergencies 32
  fever of unknown origin 79
  inflammatory markers 24, **25**
  laboratory diagnosis 1–12
  microbiological causes 134–5
  notifiable diseases 33
  opportunistic 98, 136
  parasitic 6
  prevention 136
  see also bacterial infections; fungal infections; viral infections
infection control 44, 59
  diarrhoea and vomiting 90, 91
infectivity 22
infertility 87
inflammation
  C-reactive protein 24, **25**
  temperature elevation **25**
  white cell count **25**
inflammatory bowel disease 35
inflammatory markers of infection 24, **25**
inflammatory response, neutrophils 24
influenza
  antigenic change in virus 72
  antiviral treatment 72
  complications 71
  epidemic 72
  immunisation 72
  infection control 44
  microbial investigations 71
  outbreak 72
  pandemic 72
  preventative measures 72
  travellers 74
  treatment 71
innate immunity 22–3
interferon-alpha 20
intracellular killing 23
intracranial pressure (ICP), raised 51
intravenous (iv) fluids 35
  hypotension management 54
  see also fluid resuscitation
intravenous immunoglobulin (IVIG), toxic shock syndrome 120
investigations 135
isolation of patient 28, **29**
  Clostridium difficile-associated diarrhoea 90
  tuberculosis 123
isoniazid 123, 124
itraconazole 20, 21

jaundice 45–8, 138
  drug user 138
  fever 102
  virology investigations 46

Kaposi's sarcoma 99, 100
Klebsiella
  acute pyelonephritis 116
  laboratory diagnosis 3

laboratory diagnosis of infections 1–12
  bacterial infections 1–5
  fungal 6
  technical issues 6–12
  viral infections 5–6
laboratory techniques 135
lactose fermentation 3
lamivudine 20
latex agglutination test 4, 7, 11–12, Plate 6
leg, swollen 82–5, Plate11
Legionella (legionellosis) **33**
  antigen tests 63
  laboratory diagnosis 4
  pneumonia 63
leishmaniasis, cutaneous **31**
leptospirosis 107–8
levofloxacin 18
lincosamides **16,** 18
linezolid 19
  cellulitis treatment 84
  MRSA pneumonia 43, 44
lipopolysaccharide *see* endotoxin
Listeria monocytogenes, antenatal testing **95**
lithotripsy 117–18
liver disease, hepatitis C virus 93, 95
lungs
  abscess 61, 62, 128
  cavitation 122, *123*
  consolidation 137
  fibrosis 122, *123*
  see also respiratory infections
Lyme disease
  cellulitis differential diagnosis 82, 83
  laboratory diagnosis 5

macrolides **16,** 17–18
  community-acquired pneumonia 131
macrophages, cell-mediated immunity 24
malaria
  antigen detection test 76
  blood count 76
  blood culture 76
  blood film 76, Plate 10
  chemoprophylaxis 74, 75
    in pregnancy 77
  chest X-ray 76
  creatinine level 76

diagnosis 76–7
flu-like illness 73
investigations 76
laboratory diagnosis 6
parasites 76–7
pregnancy 76, 77
travel advice 131
travel history **31**
treatment 77
urea level 76
urine culture 76
Malarone 77
malignancy, fever of unknown origin 80
measles **33,** 67
pregnancy complications **68**
measles, mumps and rubella (MMR) vaccination 67, 68, 69
postpartum immunisation 93
medications, travel history 75
mefloquine 77
meningism, definition **50**
meningitis 49, 50
adult respiratory distress syndrome 52
antibiotics 52, 54, 55, 131
bacteraemia 53–4
bacterial 50, 51–5
blood culture 52
definition **50**
imaging 52
lumbar puncture 52–3
management 52, 54, 55
notification 55
presentation 32
prognosis 55
Streptococcus pneumoniae 39
treatment 52
vaccines 55
viral 50
meningococcal sepsis 51
meningococcus
laboratory diagnosis 4
see also Neisseria meningitidis
meropenem **16,** 17
metapneumovirus 73
methicillin-resistant *Staphylococcus aureus* (MRSA) 43
antibiotic resistance mechanisms **15**
barrier nursing 28, 43
beta lactam resistance 130
hospital-acquired 43
isolation of patient 28, **29**
laboratory diagnosis 5
pneumonia 43–4, 139
metronidazole **16,** 19, 130
Clostridium difficile-associated diarrhoea 90
bacterial vaginosis treatment 86
microbiological causes of infection 134–5
microscopy
bacterial infections 1, 2
culture and antibiotic sensitivity testing (M,C&S) 1–4
viral infections 5
migraine 49, 50
minimum bactericidal concentration (MBC) 13
minimum inhibitory concentration (MIC) 13, *14*
penicillin 41
minocycline 18
mononuclear phagocytic system 23
monospot test 125
mouth infections 133
multilocus sequence typing (MLST) 9
multisystem disease, fever of unknown origin 80
myalgia 106–8
headache and fever 131
Mycobacterium tuberculosis 123
cell-mediated immunity 24
laboratory diagnosis 2
see also tuberculosis
Mycoplasma pneumoniae
laboratory diagnosis 5
pneumonia 63

nail bed, splinter haemorrhages 40
nasopharyngeal infections 133
neck stiffness 129, 137
examination 51
necrotising cellulitis 85
necrotising fasciitis 85
Neisseria gonorrhoeae
antenatal testing **95**
cervical infection 86
Neisseria meningitidis 53
immunisation after splenectomy 127
Neisseria meningitidis group C vaccine (Men C) 27, 55
neonates, varicella complications 58
neutropaenia
chemotherapy effects 112
management 113
neutrophils
chemotherapy effects 112
inflammatory response 24
night sweats 78–80
cough 122–4
fever and headache 132
weight loss and cough 130
nitric oxide (NO), endotoxic shock 26
nitric oxide synthase 26
nitrofurantoin 18–19
non-nucleoside analogue reverse transcriptase inhibitors 19
norovirus
infection control 44
outbreak 90–1
notifiable diseases 33
leptospirosis 107
measles 67
meningitis 55
rubella 67
tuberculosis 124
nucleic acid amplification test (NAAT) 7–9
Trichomonas vaginalis 87, 88
nucleic acid detection 7–9
bacterial 4–5
fungal infections 6
viral 5
nucleic acid sequencing 9
nucleoside analogue reverse transcriptase inhibitors 19
nucleoside analogues 20
nystatin 20, 86

opportunistic infections 98, 136
opsonisation 23
opsonising antibodies 127
oral rehydration therapy 89
see also fluid resuscitation
oseltamavir 20, 71
preventive treatment 72
otitis media 51
out-patient antibiotic therapy (OPAT) 110
oxygen therapy
pneumonia 62
toxic shock syndrome 120
oxytetracycline 18

pancreatitis, acute 42
pancytopaenia 112
Panton–Valentine leukocidin (PVL) 85
papilloedema 51
parainfluenzavirus 73
parasitic infections, specimen journey 6
parvovirus B19 67–9
antenatal testing **95**
pregnancy complications 68
rash 67
testing 68
Pasteurella multocida 83
pathogenicity 22
Paul–Bunnell test 125, 126
pelvic inflammatory disease (PID) 86, 87
penicillin(s) 15, **16,** 17, 130
allergy 84
pneumococcal resistance 54
rash in glandular fever 126
penicillin V 84, 85
prophylactic after splenectomy 127
penicillin-binding proteins (PBPs) 15
pericarditis, fever with pleuritic pain 61, 62
perinephric abscess 116–18
phagocytosis 22–3, 130
pharyngitis 51
piperacillin/tazobactam **16,** 17
neutropaenia 113, 114
plasmid-mediated antibiotic resistance 14
Plasmodium 76–7
laboratory diagnosis 6
Plasmodium falciparum malaria 76–7
pleural effusion 129
pleuritic pain 61–5, 128
pneumococcus *see Streptococcus pneumoniae*
Pneumocystis jirovecii
laboratory detection 6
pneumonia **98,** 99
pneumonia
antigen tests 63
atypical 5, 63
blood tests 63
chest CT 113, 114
community-acquired 62, 131
CURB-65 score 64
fever with pleuritic pain 61, 62
hospital-acquired 42–4

pneumonia *(cont.)*
ICU management 64
imaging 63, 65
influenza complication 71
MRSA 43–4, 139
neutropaenia 113
pneumococcal 129
Pneumocystis jirovecii **98,** 99
post-coronary artery bypass fever 139
serology 63
sputum culture 63
Streptococcus pneumoniae 39
travellers 74
treatment 62–3
polyenes 20
polymerase chain reaction (PCR) 8–9
pregnancy
antenatal visit 92–5
chickenpox 131
doxycycline contraindication 77
ectopic 87
hepatitis B virus 47, 48, 92, 95
hepatitis C virus 93, 95
HIV infection 93–4
malaria diagnosis 76, 77
malaria prophylaxis 77
parvovirus B19 68
rashes 67, **68**
returning traveller 75
screening tests 92
shingles contact 131
syphilis 94
varicella
complications 58
passive immunisation 59
protease inhibitors 19
Proteus, acute pyelonephritis 116, 118
pseudomembranous colitis 35
Pseudomonas, laboratory diagnosis 3
psittacosis 63
pulmonary embolism, fever with pleuritic pain 61, 62
purpura 40, 50
pyelonephritis 51
acute 115–18
pyrazinamide 123, 124
pyrexia of unknown origin (PUO or FUO) 78–80

Q fever 107
pneumonia 63
quinine 77
quinupristin–dalfopristin (QD) combination 18

rash 119–21
adolescent girl 66–9
allergy 66, 67
chickenpox 59, Plate 9
differential diagnosis 57
diffuse macular 119, 121, Plate 13
drug reaction 66, 67
history 57, 66
maculopapular 66
penicillin in glandular fever 126
petechial 40
pregnancy complications 67, **68**
purpuric 50, 129, 137
terminology 57, **58**
toxic shock syndrome 119, 121, Plate 13
vesicular 57–60
viral infection 66, 67
real-time polymerase chain reaction (RT-PCR) 8–9
red cell aplasia 68
rehydration
diarrhoea 35, 89
see also fluid resuscitation; intravenous (iv) fluids
respiratory infections
diagnostic methods 71
smoking 122
viral 20, 75
respiratory syncytial virus 73
infection control 44
laboratory detection 5, Plate 7
ribavirin therapy 20
resuscitation
ascending cholangitis 104
shock 126
retinitis, HIV-associated 96, 97, 98, 129
treatment 99, 100
Reye's syndrome, influenza complication 71
rheumatological disorders, fever of unknown origin 80
rhinovirus 73
Riamet 77
rib osteomyelitis 61, 62
ribavirin 20
rifampicin
toxic shock syndrome 120, 121
tuberculosis 123, 124
right upper quadrant pain 138
rotavirus, laboratory detection 5
rubella 67
antenatal testing 92, 93, 95
pregnancy complications **68**
susceptibility 93
testing 68, 69

Salmonella
gastroenteritis 34, 35
laboratory diagnosis 3
stool culture 36
school attendance, viral rash 67
sepsis 25
meningococcal 51
severe 25
syndrome of inappropriate antidiuretic hormone secretion 52
septic shock 25, 138
septicaemia, presentation 32
serological testing 9–12
bacterial 5
viral 6
sexually transmitted diseases 86, 87, 88, 132
shiga-like toxins 36–7
Shigella
gastroenteritis 34
laboratory diagnosis 3
stool culture 36
shingles 57–60
pregnancy contact 131
shock 25–6
ascending cholangitis 104
cardiogenic 25
causes 126
clinical management 26
endotoxic 25–6
Gram-negative 24
Gram-positive 26
hypovolaemic 25, 126
mechanisms 25–6
resuscitation 126
septic 25, 138
treatment 104
see also toxic shock syndrome (TSS)
skin
infections 85
infective endocarditis 40
lesions in HIV infection 96, 97, 99, 100
necrotising infections 85
purpura 40, 50
see also rash
slapped cheek syndrome 67
smoking 38
respiratory infections 122
social history 31
sore throat
with fever 125–7
pain on swallowing 129
spleen, ruptured 126–7
splenectomy 127
sputum culture, pneumonia 63
staphylococci
antibiotic resistance 18
bacterial endocarditis 129
coagulase-negative 110, 111, 132, 138
Staphylococcus aureus
antibiotic resistance **15**
antibiotic use 43
cellulitis 83, 84, 85
glycopeptide intermediated **15**
hospital-acquired pneumonia 42–3
laboratory diagnosis 3, Plate 4
Panton–Valentine leukocidin 85
perinephric abscess 117
toxic shock syndrome toxins 120, 121
see also methicillin-resistant *Staphylococcus aureus* (MRSA)
steroids
hoarse voice 132
meningitis treatment 52, 54
Still's disease 80
stools
cultures for diarrhoea 36
samples for parasitic infections 6
strand displacement amplification 9
streptococci
alpha-haemolytic 39
aminoglycoside resistance 14
antibiotic resistance 18
beta-haemolytic 39
fever 39
group A 44
group B 94

group C 83
group G 83
haemolysis of blood agar 3, Plate 5
Lancefield grouping 3
viridans 39
Streptococcus pneumoniae 39, 55
antibiotic resistance **15**
antigen tests 63
immunisation after splenectomy 127
laboratory diagnosis 4
meningitis 53, 54
opsonisation 23
penicillin resistance 54
pneumonia 63
Streptococcus pneumoniae vaccine (Prevenar) 27, 55
Streptococcus pyogenes
cellulitis 83, 84, 85
necrotising infections 85
sore throat 126
toxic shock syndrome toxins 120
Streptococcus sanguis 41
streptogramins 18
stridor 125
subarachnoid haemorrhage 49, 50
superantigen 22
swallowing difficulty 125, 129
syndrome of inappropriate antidiuretic hormone secretion (SIADH), sepsis 52
syphilis
antenatal screening 94
laboratory diagnosis 5
mother-to-child transmission 94
pregnancy 94
testing 46
systemic inflammatory response syndrome (SIRS) 25
score 32
tachycardia 51

T cells, cell-mediated immunity 24
tachycardia 51
tazobactam **16,** 17
teicoplanin 17, 111
cellulitis treatment 84
infected hip prosthesis 110
MRSA pneumonia 43
neutropaenia 113, 114
telbivudine 20
temperature, body
elevation in inflammation **25**
see also fever
tenofovir 20
terbinafine 21
tetracyclines **16,** 18–19
thrombocytopaenia 102, 114
thrush
oral 96, 97
vaginal 86, **87**
tigecycline 18
tonsillitis 51
toxic shock syndrome (TSS) 119–21
investigations 120
pathology 120
prognosis 120
treatment 120, 121
toxins 22
complement cascade 24
detection 4, 7
skin infections 85
see also endotoxin
toxoids 27
Toxoplasma gondii (toxoplasmosis)
antenatal screening 94
cerebral/cerebral abscesses 97, 100
fever and sore throat 125
treatment 99, 100
transient aplastic crisis (TAC) 68
transposons 14
transthoracic echocardiogram (TTE), bacterial endocarditis diagnosis 39–40
trauma, leg 82, 83
travel/travel history 31
advice on malaria 131
chemoprophylaxis 74
fever 138
flu-like illness 73–4
hepatitis 131
medications 75
pregnancy 75
return from Malawi 74–7
symptoms 74–5
Treponema pallidum, laboratory diagnosis 5
Trichomonas vaginalis 86, 87
trimethoprim **16,** 18–19
tuberculosis **33,** 122–4
case finding 124
DOTS 123
drug-resistant 123–4
extensively drug resistant 124
HIV infection 123
infection control 44
multidrug-resistant 28, **29,** 124
notifiable disease 124
pulmonary **31**
travel history **31**
treatment 123
see also Mycobacterium tuberculosis
typhoid fever **33,** 76, 77
travel history **31**

ultrasonography, abdominal for urinary tract infection 116
upper respiratory tract infections, viral 70–3
ureter, stenting 118
ureteric stone 116, 117–18
urinary frequency 137
urinary tract calculi 117–18
urinary tract infection 115–18
antenatal screening 94
recurrent 116
urine culture 116

vaccines
conjugate 27, 129
Haemophilus influenzae type b 55
hepatitis 48
inert 27
live 26–7
meningitis 55
see also immunisation; measles, mumps and rubella (MMR) vaccination
vaginal discharge 86–8
cause 86
vaginal swabs 86–7
toxic shock syndrome 121
vaginosis *see* bacterial vaginosis
valaciclovir 19
valganciclovir 19, 98
vancomycin **16,** 17
cellulitis treatment 84
Clostridium difficile-associated diarrhoea 90
Gram-negative organism resistance 14
MRSA pneumonia 43
vancomycin-resistant enterococci (VRE) **15,** 44
varicella zoster immunoglobulin (VZIG) 58, 59
varicella zoster virus
antenatal testing **95**
contacts 58–9
fetal complications 58
immunocompromised patient complications 58
infection control 44, 59
neonate complications 58
pregnancy complications 58
rash 57–60
verocytotoxic *E. coli see* enterohaemorrhagic *E. coli* (EHEC)
verotoxins 36–7
Vibrio vulnificus 84
viral infections 5–6, **12**
blood-borne in healthcare workers 48
cell-mediated immunity 24
diagnostic approaches **12**
drug therapy 19–20
rash 66, 67
specimen journey 5–6
upper respiratory tract 70–3
see also named organisms
virulence 22
voriconazole 20, 21, 114

weakness, right-sided 96
weight loss
chest infection 138
cough and night sweats 122
fever 102
night sweats and cough 130
Western blot test, HIV testing 97
white cell count
elevated 138
inflammation **25**
meningitis 53
pneumonia 64, 65
raised with fever 130

zanamivir 20
zidovudine (AZT) 19, 94
Ziehl–Neelsen stain 2, Plate 2
zoonoses 107–8